ENGLISH & AMERICAN WATCHES

English & American Watches

GEORGE DANIELS

PWP

PHILIP WILSON PUBLISHERS
Bloomsbury Publishing Plc
50 Bedford Square, London, WC1B 3DP, UK
29 Earlsfort Terrace, Dublin 2, Ireland

First published in Great Britain in 1967 by Abelard-Schuman
This reprinted edition published in 2024

A catalogue record for this book is available from the British Library
Library of Congress Cataloguing-in-Publication data has been applied for

ISBN: 978-1-7813-0131-9

2 4 6 8 10 9 7 5 3 1

Printed and bound in Dubai by Oriental Press

MIX
Paper | Supporting responsible forestry
FSC® C004800

ACKNOWLEDGEMENTS

The Authors and Publishers express their appreciation to the Collectors and Museums listed below who have either supplied photographs or made their watches available for the illustrations against their names.

Mr Sam Bloomfield: 32, 41, 47, 48.
The Trustees of the British Museum: 1, 2, 3, 4, 5, 6, 7, 8, 9, 10, 11, 12, 14, 15, 21, 23, 27, 30, 39, 40, 42, 43, 44, 45, 46.
Bulova Watch Company and Mr Eric Bruton: 50.
California Academy of Sciences: 35.
Mr Philip Coole: 38.
Hamilton Watch Company and Mr Eric Bruton: 49.
Mr Wesley R. Hauptman: 36, 37.
Mr Robert Marryat: 13, 16, 17, 18, 19, 20, 22, 24, 25, 29, 33.
Mr Ralph Shermund: 31.
Mr Peter Sbardella: 34.

The Author would also like to thank Mr Philip Coole of the British Museum for reading the manuscript and for all his help.

1967

The George Daniels Educational Trust would like to thank the private collectors of the watches that are reproduced on the front and back covers of this reprinted edition.

2024

CONTENTS

		page
	Note to Reprinted Edition	11
	Introduction	13
1	*The Beginning*	15
2	*English Watchmaking in the Seventeenth Century*	18
3	*The Balance Spring*	21
4	*Pioneer Precision Work*	26
5	*The Emergence of the Precision Watch*	33
6	*The Early Lever and Duplex Escapements*	37
7	*The English Watch: Final Development and Decline*	41
8	*American Watchmaking*	46
9	*Post-War Developments in America*	55
Appendix A	*The Verge*	59
Appendix B	*Cylinder Escapement*	61
Appendix C	*The Duplex*	63
Appendix D	*The Chronometer*	65
Appendix E	*The Lever*	67
	Biographical Notes	69
	Plates	77
	Index	124
	Bibliography	128

HALF TONE ILLUSTRATIONS

		page
1	*Jacques Bulke, c.*1600	79
2	*Hugh Roberts, c.*1620	81
3	*Henry Terold, c.*1640	81
4	*Robert Grinkin, c.*1640	83
5	*Edward Clough, c.*1640	83
6	*Thomas Tompion, c.*1680	85
7	*Ahasuerus Fromanteel, c.*1695	85
8	*John Bushman, c.*1695	87
9	*Peter Fardoil, c.*1700	87
10	*Joseph Banks, c.*1700	87
11	*William Peacock, c.*1700	89
12	*George Graham, c.*1735	89
13	*Larcum Kendal, h.m.* 1777	91
14	*Thomas Martin, c.*1778	91
15	*John Arnold No.* 37, *h.m.* 1779	93
16	*John Arnold No.* 9/61, *h.m.* 1789	93
17	*John Arnold No.* 109/410, *h.m.* 1784	93
18	*John Arnold No.* 145, *c.*1800	93
19	*Peter Litherland, h.m.* 1794	95
20	*Mudge & Dutton, h.m.* 1800	95
21	*Thomas Earnshaw, c.*1798	97
22	*Thomas Earnshaw, c.*1800	97
23	*Edward Massey, h.m.* 1819	99
24	*William Ilbery, c.*1820	99
25	*J. R. Arnold, h.m.* 1820	101
26	*J. F. Cole,* 1830	101
27	*Thomas Penkett, h.m.* 1832	103

		page
28	*Parkinson & Frodsham, c.*1860	103
29	*W. M. Godfrey, h.m.* 1870	105
30	*John Bull (Lancashire Watch Co.), c.*1905	105
31	*Charles Frodsham, c.*1910	107
32	*Nicole Nielson, c.*1913	107
33	*Rotherham, c.*1923	109
34	*John Harwood, c.*1930	109
35	*Luther Goddard, c.*1816	111
36	*Dennison, Howard & Davis, c.*1854	111
37	*E. Howard, c.*1858	113
38	*E. Howard, c.*1865	113
39	*Charles Fasoldt, c.*1865	115
40	*Auburndale Rotary, c.*1876	115
41	*Albert Potter, c.*1877	117
42	*Waterbury 'A', c.*1878	117
43	*A. Waterbury 'F', c.*1888	117
44	*Waterbury 'I', 'The Trump', c.*1889	119
45	*Waltham, c.*1890	119
46	*Ingersoll (manufactured by the New Haven Clock Co.), c.*1892	121
47	*Waltham Premier Maximus, c.*1909	121
48	*E. Howard, c.*1909	123
49	*Hamilton,* 1966	123
50	*Bulova, Accutron,* 1966	123

LINE ILLUSTRATIONS

		page
Fig. 1	*The verge escapement*	60
Fig. 2	*The cylinder escapement*	62
Fig. 3	*The duplex escapement*	64
Fig. 4	*Arnold's chronometer escapement*	66
Fig. 5	*Earnshaw's chronometer escapement*	66
Fig. 6	*Lever, English ratchet tooth escapement*	68
Fig. 7	*Lever, club tooth escapement*	68

Note to Reprinted Edition

This is the second book written by the late Dr George Daniels CBE FBHI who was widely regarded as the greatest watchmaker in the twentieth century. The important contribution made by England and America in the technical development of the pocket watch had not been previously surveyed.

The book was first published by Abelard-Schuman in 1967 two years after the publication of *Watches: A Complete History of the Technical and Decorative Development of the Watch* that Daniels had written with his co-author Cecil (Sam) Clutton to great acclaim.

Although we have been unable to source new photography, our wish is to keep George's publications in print as they remain invaluable reference sources for the watch, clockmakers and collectors of today.

The George Daniels Educational Trust
2024

Introduction

THE RELIABILITY AND ACCURACY OF THE MODERN WATCH IS, BY MOST wearers, accepted without question. Even the cheapest watches are capable of keeping time to within a minute a day and more sophisticated examples to within a few seconds a day.

There is very little in the most modern watches to suggest their ancient origins but this near perfection in time keeping is the result of centuries of invention and experiment by horologists both professional and amateur. Some of their inventions were of such fundamental importance that they continue to be used in the modern watch with only detail alteration. Others, following different principles, met with less success, but even so their work often indicated the true direction for the next generation to follow.

The development, and often invention, in England of the fundamental principles of precision watches and their early mass production in America, where new ideas were more readily accepted, combined to show the direction and method that the modern industry was to follow.

Although it would be incorrect to suggest that the development of the watch was confined to the English and American industries, these two countries have in fact made the most direct contribution. The vision of the American industry has indeed been responsible for the greatest advance in modern technique with the introduction of electricity to mass produced watches.

It is for this reason that this book confines itself to the products and makers of these two countries, although it must be clearly understood that fine watches were being produced by many celebrated makers on the continent of Europe.

The newcomer to the subject is likely to find himself overwhelmed by the mass of facts already available and to find a simple basic history more useful as a foundation for future studies. This book is directed particularly at this steadily growing number of people but at the same time continues the study of the watch to include the latest developments which will be of interest to the more knowledgeable enthusiast. More particularly may this be so of the photographs, the majority of which have not before been published.

Confusing discussion of the technical features of the various contributing inventions is avoided. Their importance and merits are nevertheless clearly underlined. The dates of the inventions are included in the text in chronological order. The dates (where known) of the inventors and makers are confined to the biographical notes. These notes are concerned only with the men and watchmakers mentioned in the text and represent but a fraction of the number of known horologists and manufacturers.

The photographs of cases and movements are chosen to illustrate the change in external appearance that accompanied the development of the movement.

So that the reader may clearly understand the function of the escapements commonly used in the watches, seven line drawings are included with accompanying explanations.

1. *The Beginning*

THE EARLIEST SURVIVING EXAMPLES OF MECHANICAL TIMEKEEPING IN England are to be found in weight driven clocks made in the last quarter of the fourteenth century. There is ample evidence to suggest that they had their origins in Italy some fifty years earlier, with Germany and France in the field soon after, if not initially. Continuous refinement of the mechanism resulted in smaller clocks, so that, by the middle of the fifteenth century, domestic clocks were by no means uncommon in these countries. Being weight driven, their use was confined to one room in which they would usually be fixed to the wall at a height sufficient to allow for the fall of the weights. The inconvenience of this arrangement prompted makers to look for alternative means of driving the wheels, and the answer was found in the coiled ribbon of steel wound round the arbor which previously carried the ropes for the weights.

The earliest evidence of a spring driven clock indicates that the credit for this invention must go to Burgundy or Flanders, but it was in Germany that the most important development took place. In Nuremburg, a great centre for metal-working, Peter Henlein made small portable clocks. Before the end of the first quarter of the sixteenth century these had been so refined as to make it quite convenient to wear them on a cord strung around the neck or waist. Spherical in shape, these first watches, known as musk apples, were some three inches in diameter and included a striking mechanism. The cases of gilt metal were decorated by engraving the surface. Drum shaped cases very quickly followed and soon superseded the spherical type.

In France watchmaking kept pace with Germany and in both countries development was confined to shaping the case in order to make it more convenient to wear. A lid was fitted to protect the hand and this, in the German products, was pierced to enable the wearer to see the time without opening the watch. In construction and decoration the French watch was in every way more refined than the German counterpart, which always retained the coarseness of the earlier portable clocks.

The most fundamental difference of approach was to be found in the movement which, in the German product, was constructed of iron, the French being made, except for the pinions, entirely of brass.

It is well known that a spring exerts its greatest force when fully wound and that this force decreases proportionately with unwinding. To equalize the force so that the escapement receives a more constant flow of power the German watches mostly employed the extremely crude stackfreed. This in its simplest form consists of a snail-shaped cam fitted to the mainspring arbor. Fixed at one end to the plate of the movement is a powerful spring. The free end of the spring carries a small roller which presses against the cam. When fully wound the roller acts against the largest radius of the cam which, while the movement is in action, slowly rotates on its arbor and presents a constantly decreasing radius to the roller. Thus it relieves itself of the pressure of the spring and the braking effect on the power of the mainspring. This unsatisfactory device appears to be completely German in origin and has never appeared on a watch made outside Germany.

The French preferred the much more sophisticated fusee and cord. Principally it consists of a barrel containing the mainspring. The inner end of the spring is hooked to the arbor upon which the barrel rotates. The driving wheel of the watch carries on its arbor a spirally grooved pulley of continuously decreasing diameter. The barrel and pulley are joined by a cord which, in the unwound state of the mainspring, is wound round the barrel. Winding the watch transfers the cord from the barrel to the pulley and at the completion of winding the cord is pulling at the smallest diameter of the pulley. During unwinding the cord is wound back onto the barrel and the fall in power of the spring is assisted by the increasing diameter of the pulley. The origins of this extremely ingenious device are obscure but the evidence is that it was used from the earliest days of spring driven clocks, and is still in use today for marine chronometers.

By the third quarter of the sixteenth century brass plates appear in German watches, but the stackfreed continued to be used. The standard of wheel dividing was inferior to the French, and the decoration and finish continued to be coarse. The French centre at Blois grew rapidly and the industry spread quickly to such places as Rouen, Paris, Dijon and Lyons, etc., and by the end of the century completely eclipsed the German industry, which went into a decline. Watches continued to be made in Germany but in relatively small numbers. As with the products of the few other European countries producing watches, they were dominated by the French industry, and played no further part in the early development of the watch.

2. *English Watchmaking, Seventeenth Century*

EXAMPLES OF ENGLISH WATCHES MADE BEFORE THE END OF THE sixteenth century are rare and records of English makers do not occur before 1580. The surviving watches show French influence in the movements and both French and German influence in the case style and decoration. Thus a watch would have a brass movement with fusee and floreat balance cock, and a case in the German style with curved edges and a lid with apertures for the figures of the hours.

During the first quarter of the seventeenth century the English industry grew and settled into a native style closely allied to the French for both movement and case. A feature peculiar to English watches was the decorated border to the back plate, but this was abandoned by 1630. The mechanisms were sometimes quite complicated, and included striking the hours and an alarm. The dials also became more complicated and such information as the date, age and phase of the moon, and the sign and degree of the zodiac were often included.

However, the continuous process of refinement in decoration and the variety of complicated devices contributed nothing towards correcting the biggest failing of the watch. In spite of these innovations it was quite incapable of indicating the passing hours with any sort of regularity. Its rate might vary by as much as a half hour in twenty-four and, owing to irregularities in the wheel dividing, it would often stop when worn. It was probably for this reason that the three wheel train, necessitating winding twice a day, was retained for so long until, in the second half of the seventeenth century, improved methods of dividing made the four wheel train a practical proposition.

Apart from refinement in finish and decoration, at this stage there was nothing the watchmaker could do to improve the movement. An inordinate amount of labour was expended in fashioning the plate pillars which, at their most fanciful, were beautifully and minutely detailed. In order to enable the bemused wearer to regulate the rate to some degree, the set up from the mainspring was secured by a worm and wheel. The tension of the spring could then be varied by the application of a key to the worm shaft. The single hand was driven from a pinion filed into the end of the fusee arbor. This arrangement necessitated the fusee arbor being made in two halves so that the watch could be wound without turning the hand backwards. The system employed required that the great wheel be mounted or driven on to an arbor, one end of which carried the pinion for driving the hand and the other end running freely in the fusee, the two being coupled in the driven direction by ratchet and pawl. This rather feeble method resulted in the fusee cone rocking on the great wheel and forcing itself towards the plate, thus further absorbing some of the spring power.

The balance cock was secured by a screw and steady pin instead of the earlier method of fitting to a square peg and pinning through. This ensured a more positive re-location after dismantling. During the whole of the seventeenth century the cock grew larger and stronger and afforded greater protection to the balance. During the middle of the seventeenth century the balance cock, the regulator worm brackets and plate pillars were all carefully detailed and fashioned and are a particular feature of watches of this period.

If the watchmaker could not effect any improvement in the movement he made the most valiant efforts to distract the customer's attention from its shortcomings by the lavish and skilful decoration of the case.

With the beginning of the English school, cases followed the style of France and Germany and were usually of metal engraved and gilt. More fanciful shapes soon followed, oval and octagonal being very popular. Cases carved from rock crystal polished and faceted are particularly beautiful, enabling the movement to be seen through the case.

Gold and silver cases appeared in the form of flowers, books, skulls and crosses. Nearly all are most beautifully engraved with pastoral or classical scenes, the cruciforms being usually engraved with religious scenes. In contrast to these were the so-called 'puritan' watches. Completely smooth and usually about the size of a pigeon's egg, they are devoid of external decoration.

Probably inspired by the crystal cases, glasses began to appear. In the earliest examples the advantage of being able to see the time without opening the front was nullified by enclosing the watch in a separate outer case. Later these too had an aperture.

By the third quarter of the century all watches had an outer case and, while some inner cases were still decorated, the outer was frequently lavishly worked. These were covered with leather, tortoiseshell, or fish skin and decorated with pins arranged in scroll or floreat patterns. The dials of silver or gilt metal were engraved inside the chapter ring with pastoral or religious scenes, the outer zone engraved with a floreat design. Sometimes the zones would be painted enamel, but this is extremely rare. In the second quarter of the century the chapter ring was often of engraved silver applied to a gilt background. The single blue steel hand had a long tail to assist in setting.

In sixty years the watch had developed into a very pretty object, but its chief interest lay in the novelty of its appearance, for its timekeeping was no better than that of the earliest spring driven clocks. The last quarter of the century was to see a significant change.

3. *The Balance Spring*

FROM THE BEGINNING OF MECHANICAL TIMEKEEPING UNTIL THE FIRST quarter of the eighteenth century the verge escapement [Fig. 1] was used exclusively for both clocks and watches, there being no alternative mechanism. In this escapement the period of oscillation of the balance is dependent upon the power delivered by the 'scape or crown wheel. The poor quality mainspring, and inferior wheel trains combined to effect an inconstant impulse to the balances thus producing a proportionately erratic period of oscillation.

Depending for its motive power on a falling weight, the clock had a distinct advantage, but even so the variation in rate might be as much as five minutes in one day. Leonardo da Vinci was the first to suggest that a pendulum be used as a means of controlling the escapement in clocks, and the first practical experiments with the pendulum were conducted by Galileo. Seven years after his death in 1642 his son Vincenzio began the construction of a clock to Galileo's plan. Vincenzio himself died before completing the clock but the proof of its existence is to be found in the drawing preserved in the Medicean Archives at Florence. It is also mentioned in the inventory of the belongings of Vincenzio's widow who died in 1669.

In 1657 the celebrated Dutch astronomer and mathematician, Christian Huygens, ordered a clock to be constructed embodying his own ideas for the use of a pendulum. The result was a startling success and the timekeeping of clocks improved to such an extent that the variation amounted only to seconds per day. In direct contrast, watches were still exceedingly unreliable. As it was now obvious that close timekeeping was at last a possibility, the watchmaker was faced with the necessity of swiftly improving both his product and his reputation.

In 1662, in London, the scientifically minded Charles II founded the Royal Society. Its purpose was, and still is, to further and publicize the advancement of scientific knowledge. Robert Hooke, professor of geometry at Gresham College, was appointed curator of experiments. Although Hooke was primarily interested in microscopes his enquiring mind ranged over most of the scientific problems of his day. One such interest was the improvement of balance wheel timekeepers. His experiments in this field were directed to applying a spring controller to the balance. By 1668 he had been experimenting for ten years without any particular inspiration or success. It is probable that he interested himself in too great a variety of subjects so that his researches were somewhat haphazard. The unenlightened approach on the part of the watchmaker to the practical application of his experiments would no doubt irritate a man of his none too patient disposition, even to the extent of disrupting his judgment on his own conclusions. It was not until 1675, when Christian Huygens announced his invention of the spiral spring that Hooke remembered that he had already made the same discovery. A watch commissioned by Hooke from Thomas Tompion in that year as a counter claim to Huygens' invention would not seem to have included a spiral spring in its specification. This may not be of significance, since Hooke may have hurriedly decided to seek an alternative and possibly better arrangement, the more effectively to dismiss Huygens' claims.

However, no patent was granted by the King and, as with the invention of the pendulum, Huygens viewed rival claimants with equanimity. Certainly it would seem that Huygens was intellectually the better equipped to solve the problem, but there is nevertheless quite strong evidence to support Hooke's claims.

With the invention of the spiral spring came the greatest step forward in the history of horology. English makers were quick to master the technique of its application, and with the improved trains produced with Hooke's very successful dividing engine, invented in 1670, English watches were supreme. By minute division of labour the watchmaker was able to buy in the rough almost any part used in the watch. These, by further sub-division of labour, were finished and fitted to the plates, which could also be bought in the rough state.

Constructed in this way and fitted with the new balance spring they were capable of keeping correct time to within two or three minutes per day. The worm and wheel regulator, being no longer required, was abandoned, and the regulation was achieved by sliding a curb along the spring so as to lengthen or shorten it effectively as required. The extra power required to overcome the resistance of the spring was achieved by increasing the height of the mainspring, necessitating a change from gut to chain for the fusee. Watches therefore became thicker and bulkier.

No longer needing to justify his existence, the maker devoted more attention to the movement and less to the case. These became simpler and were often merely covered in black leather or fish skin dyed green. Some were without decoration of any kind, being plain gold or silver.

The improvement in the gear train had already been responsible for the introduction, in some instances, of the four wheel train, which by 1685 was universal, so that watches required winding only once each day.

The improved timekeeping through the use of the balance spring now made it possible to indicate minutes, and this led to a rearrangement of the train. The extra wheel was planted in the middle of the watch and geared to turn once an hour. Its extended front pivot carried the minute hand, and by suitable gearing the hour hand was made to revolve concentrically with it once in twelve hours.

As it was no longer necessary to drive the hands from the fusee arbor, the great wheel was fitted freely to the arbor, with a collar to secure it. By this means the two were held together and eliminated the pressure caused by the fusee rocking. As a result the train ran more smoothly and power fluctuations were reduced.

The dials were suitably engraved with Roman numerals to show the hours and surrounded by minute divisions, with the five minute divisions engraved in Arabic numerals. A seconds dial was sometimes included, but examples are rare.

After the exciting improvement effected by the balance spring interest in the movement waned, and once again the watchmaker turned his attentions to the appearance of the watch. The novelty of minute indications proved irresistible as a means of embellishment, and this emphasis on decoration continued unabated, until the end of the first quarter of the eighteenth century brought improvement in the mechanism of the movement.

The most practical of these unnecessary complications is the differential system. The minutes are marked in the conventional manner around the edge of the dial and indicated by the usual minute hand. The hours are engraved on a separate disc within the minute circle and revolve in the ratio of 1 – 1/12 with the minute hand. Thus the single hand indicates both hours and minutes.

The six hour dial employs two sets of numerals. The numbers 7 – 12 in Arabic are superimposed upon the numbers I – VI in Roman. The six divisions of the dial are equally divided into thirty sub-divisions of two minutes. The single hand turning once in six hours indicates both minutes and hours, provided there is no confusion about which set of figures to read.

The revolving hour system also cramps the minute divisions by disposing them around half the circumference of the dial. Within these divisions is a semi-circular slot behind which the hour revolves, indicating the minutes in passing. At one end of the slot the hour disappears and at the other end the next hour appears. Behind the blank section of the dial the discs, upon which the hour numerals are engraved, are turned by a pin to the following hour, which in its turn revolves past the minute divisions.

In the sun and moon dial the whole of the circumference of the dial is used for the minute divisions, these being indicated by the conventional minute hand. Within their circle the hours are disposed in a semicircle. Starting from the left of the dial, at the 45 minute mark, they number 6 – 12 to the pendant, and 1– 6 up to the 15 minute mark. The whole of the dial within the semicircle of hours is cut away, and revolving behind this aperture once in twenty-four hours is a disc. One half of the disc is blue, with a moon to indicate the hour in passing, the other half is silver, with a sun to indicate the daylight hours.

None of these dials is common and each one is, in its individual way, an attractive novelty. But for sheer lack of artistic and mechanical inspiration the prize must go to the expanding minute hand. In this arrangement the hours are set in a circle close to the centre of the dial, with a very short hour hand. Concentric with it is the minute hand which indicates the minutes on an oval zone. The minute hand grows and shrinks in length alternately to remain within the oval minute zone. To contain the cam for producing this effect the hand must be extremely thick and wide and is correspondingly clumsy.

To emphasize the improved relationship in timekeeping between the pendulum clock and the balance spring watch, a mock pendulum was sometimes included on the watch dial. To achieve this the movement was in effect turned upside down and the dial fitted to the back plate. Through an aperture in the dial an oscillating spoke of the balance carrying a mock pendulum bob could be seen.

Thus did the late seventeenth and early eighteenth century watchmaker occupy himself without making any attempt to understand the science of his art or effect any mechanical improvement to enhance the balance spring.

The great makers of the period with sufficient income to enable them to study, and possibly correct some of the failings of their products, preferred to devote their energies to increasing their output. The results are beautiful and interesting but without inspiration. It was a period of complacent stagnation.

4. *Pioneer Precision Work*

IN 1726 THE INTRODUCTION OF THE CYLINDER ESCAPEMENT [FIG. 2] by George Graham brought an end to this unprogressive period. The new escapement, by eliminating the recoil of the verge, was better able to absorb the variations of impulse delivered by the train. The original idea for this escapement is generally ascribed to Tompion, with whom Graham was for a short time in partnership. Whatever the circumstances it is certain that Graham's more enlightened approach was entirely responsible for its success. His appreciation of the problems of achieving closer timekeeping and his direct and intelligent approach to their solution are clearly illustrated in his application of mercurial temperature compensation, and the elimination of the recoil in pendulum clocks. The advantage of the latter principle is the foundation of the cylinder escapement, and one of which Graham was obviously very conscious. The so-called Tompion cylinder impulsed in one direction only and was more suited to clocks, in which form it was called a 'tick-tack', a description which requires no explanation.

The performance of the Graham escapement is superior to the verge, so that timekeeping to a variation of a minute a day in average temperatures is possible. This soon deteriorates with want of cleaning. In this respect the verge had a slight advantage. The 'scape wheel with its impulse planes raised on stalks above the face of the wheel was not easy to make. This difficulty was soon overcome by the construction of suitable machinery, and Graham devoted the remainder of his life to producing watches using this escapement. He seems to have taken no further interest in the science of horology but rested contentedly on his well deserved laurels. This is a pity for although his escapement was reasonably successful it was very badly proportioned. The cylinder was so large in diameter as to cause excessive friction during the supplementary arc, and the extra power required to overcome this fault caused fairly rapid wear.

Graham's eminence during the second quarter of the eighteenth century enabled him to set the fashion in appearance. Enamel dials in the same style, but more elegantly proportioned than the earlier *champlevé* dials, became common, and concentric hour and minute hands were established. Later examples had also a centre seconds hand that could, by a lever arresting the balance or train, be stopped and, sometimes with difficulty, started again at will. Repeating mechanism invented simultaneously in 1686 by Edward Barlow and Daniel Quare was often included. Depressing the pendant wound the mechanism and caused the hours and quarters to be struck on a bell screwed to the inside of the case. The hammers worked through slots cut in the edge of the brass gilt dust cover which became a standard and easily detachable fitting to the movement.

In 1704 Facio and Debaufre had discovered the art of piercing jewelled bearings, and they were used for the balance pivots, in conjunction with ruby or diamond end stones. Their use later spread to the 'scape wheel holes and through successive makers of the best watches to all the holes. By 1775, in John Arnold's[1] watches, even the fusee is jewelled.

Although there is no great advantage to be had from jewelling so far back in the train the advantage of jewelling the quicker moving parts is considerable. Provided oil is present, wear of the pivots in jewelled holes is eliminated and thus a more stable condition for maintaining uniformity of rate is achieved. The polished face of the jewel assists in keeping the oil at the pivot, whereas with the hole direct in the brass plate it is more likely to spread over the plate away from the pivot. Dirt falling into the holes and mixing with the oil will not bed into the surface and score the pivot as would be the case with brass. The ability to jewel with such facility gave England a great ascendancy over the French, always her greatest rivals in horology. So skilled in the art were the English watchmakers that, by 1775, even the cylinders for Graham's escapement were fairly commonly

[1]See p. 33

made from ruby, whereas the French did not fully master the art for a further twenty years.

The pair-case outer, undecorated or covered with leather and pin work, increasingly gave way to *repoussé* decoration in which classical and mythological scenes were raised on the surface by hammering from inside.

These watches set a firm fashion and the celebrated Thomas Mudge, seeing no way of improving them, after Graham's death in 1751, manufactured identical watches in great number and founded the greater part of his business upon them.

During the period in which Graham was experimenting with his mercurial pendulum the Yorkshireman John Harrison, together with his elder brother James, was also investigating on the same lines, but using the dissimilar expansion of steel and brass as a means of maintaining the length of a pendulum constant. He could have had no knowledge of Graham's experiments since he had had no contact with horologists and had never been to London. His work is the more remarkable because he could have been no more than twenty-five years of age when conducting these experiments and his knowledge of arithmetic and the use of his tools were almost entirely self taught.

In 1728 Harrison came to London with his compensating pendulum, and in addition brought with him the curious frictionless 'grasshopper' escapement, so-called because of its resemblance to that insect. This was his brother's invention.

The purpose of his visit was to raise money to enable him to construct a timekeeper capable of determining longitude, for which, in 1714, a prize of £20,000 had been offered by an Act of Queen Anne.

To this end he laboured for the remainder of his life and eventually succeeded in producing a machine that met the requirements of the Act and, after ten years of wrangling with the Board of Longitude, secured the £20,000 reward.

Although his prize winning machine was in appearance like a very large watch, he did not concern himself with the construction of watches. Indeed his second attempt at a marine timepiece, completed in 1739, weighed no less than one hundred and seventy pounds with its box and fittings.

This remarkably clever man had a profound influence on the future trend of horology, although, except for the device invented to prevent reversal of the train during the winding, none of his ideas survived the search he started for a precision timekeeper.

This latter device, aptly named the 'maintaining power' after its function, employs an extra ratchet wheel sandwiched between the fusee cone and the great wheel. The pawl of the great wheel is transferred to the ratchet wheel, and the power from the fusee delivered to the great wheel via a semi-circular spring coupling it with the ratchet wheel. With reversal of the fusee during winding a pawl, pivoted between the plates, engages the ratchet wheel and prevents it reversing with the fusee. Thus the semi-circular spring is kept in tension and maintains the working of the watch during winding. Harrison's greatest contribution to the advancement of the science was the vision he left with his successors of the practical possibility of accurate timekeeping.[1]

In 1765 close study of Harrison's work had the considerable effect of prompting Thomas Mudge to leave his factory in the capable hands of others and devote most of his time to producing a machine that would exceed the performance of Harrison's. The escapement of Harrison's final machine was in essentials the verge found in common watches. The pallets were jewelled and specially shaped to resist the effects of fluctuations in power. Having satisfied himself that nothing more could be done with the escapement, he then eliminated the power fluctuations by driving the 'scape wheel with a separate small spring wound every seven and a half seconds by the mainspring through the normal train. Temperature correction was secured by mounting the regulator curb pins at the end of a lamination comprized of a strip of steel and a strip of brass riveted together with the brass side nearest the spring stud. Fixed to the movement plate at one end, with the curb pins embracing the balance spring at the free end, a rise in temperature caused the strip to curl away from the stud and thus effectively shorten the spring to compensate for the loss of elasticity. Conversely, for a fall in temperature the curb allowed an increase in the free length of the spring.

The compensation curb Mudge copied freely but for the escapement he substituted his own complex and ingenious arrangement. In this the 'scape wheel became merely an agent for winding the impulse springs which, at the dictation

[1] A complete and remarkable account of his endeavours is to be found in *The Marine Chronometer* by Lieutenant-Commander Rupert T. Gould.

of the balance, impulsed it at each vibration. This escapement, although too complex to succeed fully, for there were few who could command sufficient skill to construct it, did contain one important principle upon which future success was partly to depend. This was the freedom of the balance from the disturbing influence of the train after impulse had been delivered. An important step forward, this ensured the further successful development of the watch escapement. Mudge finished his first machine in 1774, and it proved relatively successful. It was in fact a marine timepiece and much too delicate and complex to be reduced to the size of a watch.

The necessity for the maximum free vibration of the balance had been made clear by the Frenchman Pierre Le Roy eight years earlier. The marine timekeeper employing this principle and completed by him in 1766 employed also a new principle for the correction of the temperature errors. This was later to be adopted, to the exclusion of all others, for close timekeeping. It consisted of a device mounted on the balance wheel to effect, with change of temperature, a change in the radius of gyration. Thus with a decrease in temperature the inertia of the balance was increased, and correspondingly reduced for a rise in temperature.

Of these two principles, upon which the future supremacy of English watch-making was to be founded, Mudge was very conscious of the importance of the first. He had made, in 1769, a remarkable watch based upon the escapement invented in 1722 by the Abbé d'Hautefeuille. This escapement, resembling Graham's escapement for pendulum clocks, carried upon the pallet arbor a segmental rack. Formed to a curve struck from the pallet arbor and extending horizontally to act as a lever, the rack engaged a pinion on the balance arbor and so communicated the impulse from the 'scape wheel. No doubt in this form, even at so early a date, it had a lively and convincing action.

Being much taken with the necessity for freeing the balance from all but the interference of unlocking and impulse, Mudge devised a method of unshackling the balance from the lever. This he achieved by dispensing with the rack and terminating the lever with a two pronged fork. The pinion on the balance arbor he replaced with two cams arranged one above the other, their profiles to engage the lever prongs arranged in corresponding elevation. Thus after the unlocking and impulse were completed the balance was free to continue its vibration unhampered by the mechanism that supplied its vitality. To prevent the lever accidentally unlocking, and so finding itself on the wrong side of the advancing cam, a small pin fixed to the lever banked against a slotted roller on the balance arbor. During the engagement of the lever and balance the pin passed into the slot.

Alas, poor Mudge. His idea was sound enough, indeed it was subsequently to prove more influential than he ever imagined. As to his execution of his invention, he exhibited an obtuseness of thought and lack of understanding of the principles involved such as to suggest that he was not at all sure what he was trying to do. Even allowing for the unprogressive state of the science of horology at the time, some lessons in proportion could be learned from the verge and cylinder escapements. This particular watch, although capable of better timekeeping than any other made before it, was a disappointment to Mudge. In wear it was inclined to stop and was an erratic timekeeper. He seems to have convinced himself that the fault lay in the temperature compensation and, in a letter written in 1773, 'a species of error that I had not before the least notion of'. The fact is that the escapement is so badly proportioned and the balance so large and heavy that the relatively puny mainspring has insufficient power to overcome the frictional losses. The species of error referred to was most probably escapement error exaggerated by the excessive escaping angle of about forty-five degrees and the small supplementary arc of the same magnitude. With so little vitality the safe banking of the lever was uncertain and this would give rise to further considerable variation during wear through the impulse cams fouling the tips of the prongs of the lever. Mudge soon lost interest in this watch, for he had intended it to be a close timekeeper for wear on the person. When it failed in this respect he turned his attention to perfecting his marine timekeepers, although as a last fling he insisted that, as a static timekeeper, it was at least the equal of John Arnold's, whose early work was just then appearing.

It is worth noting that in a letter written in 1774 Mudge referred to a variation caused by the dampness of the air. 'I have constantly observed,' he writes, 'when the air has been very damp, without any difference worth notice in either barometer or thermometer, the watch has gone too slow . . .' This is interesting because towards the end of the nineteenth century the firm of Le Roy et Cie discovered, during experiments with precision regulators, that if unoiled at the pallets the rate varied with the humidity of the air.

In his watch Mudge included an isochronal pin to equalize the rates for change of arc. This was simply an adjustable pin against which the balance spring banked during its contracting vibration. The pin was adjustable so that, according to the arc of the watch, the spring would be effectively shortened by an amount necessary to quicken the small arcs which are, for the lever escapement, always slow. The pin, being adjusted to suit the natural error would, if the error changed, be unable to compensate. It is possible that Mudge did not intend the escapement to be oiled and, if this is so, an increase in humidity would bring about a change of natural error and consequent relative loss in the long arcs. Later constructors of the escapement intended that it should be oiled and made provision for oil retention in the 'scape wheel teeth. There being no further reference to the error it was thus, presumably, eliminated.

Although Mudge achieved no great success with his watch its construction was nevertheless a milestone in progressive development. The principles of scientific watchmaking which it incorporated were to be the basis for the future worldwide watch manufacturing industry. It is worth noting also, that at the time of its construction there were probably only half a dozen men in the world who could have undertaken the task, and fewer who could have equalled the perfection of finish which was a feature of Mudge's work. For his finish he was justly famed in his own time. In the following thirty years his high standards were to become general throughout the English industry and make their contribution to English supremacy.

5. *The Emergence of the Precision Watch*

THE END OF THE THIRD QUARTER OF THE EIGHTEENTH CENTURY showed, in spite of Harrison's and Mudge's discoveries, no improvement in the common watch. It was almost identical in movement and case to the watches produced by Graham fifty years earlier. Pair cases, often quite plain, were universal. Decoration was confined mainly to enamel plaques with heraldic or symbolic themes. It was a period of unimaginative repetition with the ordinary watchmaker apparently uninspired by the examples of Harrison and Mudge.

One man alone distinguished himself. John Arnold, born at Bodmin in Cornwall, had, by 1764, established himself in London as a watchmaker. In this year he presented to George III a watch with ruby cylinder escapement and repeating mechanism. Its size was only a little over half an inch in diameter. Thus he showed his ability as a craftsman; he was soon to demonstrate that this was combined with a rare intelligence for the appreciation of the correct principles of his art.

In 1770 he completed his first attempt at a marine timekeeper. For the temperature compensation he used Harrison's bimetallic curb. The escapement was detached and of his own design. In it the 'scape wheel delivered the impulse to a pallet on the balance arbor. The wheel was then arrested by a pivoted lever lifted vertically into the path of the teeth by an inclined plane attached to the arbor below the impulse pallet. Two more machines quickly followed and all were tried at sea. Their performance was disappointing and Arnold was reproved by the Board of Longitude for his failures. Stung by their rejection he worked at his ideas for improvement, and in 1776 patented the helical balance spring and a balance containing a device for temperature compensation. In the previous year John Harrison had suggested this alternative to the compensation curb, as also had Le Roy in France. The escapement also was redesigned and the pivoted lever arranged to move horizontally to arrest the wheel. The inclined plane on the balance arbor was dispensed with and the lever returned by a spring. In this form it was a great improvement on its predecessors. Indeed so well did it perform and so simple was its construction that Arnold produced some small enough to wear in the pocket. These proving successful, he sent one to Greenwich for trial, and after being tested in wear for over twelve months its greatest daily error was only three seconds. A truly remarkable performance when it is considered that only ten years earlier the best a watch could do, under favourable conditions, was in the order of an error of a minute a day.

Working continuously to improve on the performance of both marine and pocket pieces Arnold, within twelve months, patented an improved balance. In this the bimetallic strips were lengthened to increase the movement at the free end and the weights which they carried were reduced in size. This gave a greater ratio of movement for a given change of inertia and proportionately reduced the errors. With the earlier balance the strips were fixed by their ends at equidistant chords of the balance. A change in temperature caused the strips to bow and carry with them a weight at the edge of the balance and fixed to the strip by a rod passing through the balance rim. The stress introduced in the strip by fixing to the rim would cause a variation in the extent of the compensation. In the new balance this was eliminated by fixing one end only of the strip and attaching the weight to the free end.

At the same time Arnold improved his helical balance spring by curving the ends to a radius smaller than the main body of the spring. It is a fact that if a length of spring is fixed at one end and then wound into a circle the point of least curvature will occur at the fixing. Thus in a balance spring the point of least flexion will be at the fixing. It therefore follows that the point of greatest flexion will be diametrically opposite the fixing. The movement of the spring consequently is not equally disposed about the centre of the balance arbor. The effect of this is to cause side thrust on the balance pivots and, more serious, the point of attachment to the balance.

In curving the ends of the spring to reduce the effects of the side thrust Arnold perceived that, by manipulation of the extent of the curve, the thrust could be adjusted to overcome the individual escapement error of each machine. In this discovery he showed his genius for original thought. He possessed a sense of mechanical principles and functional proportions that enabled him to raise his craft to a high degree of art, and combined with his scientific approach to the requirements and adjustment of his timekeepers it sets him apart from all his predecessors. Even Le Roy who understood the requirements was unable to interpret them as clearly as Arnold.

Variations in the rate caused by the pivots of the lever or detent, as it is usually called, were responsible for the redesign of the escapement. In the new plan [Fig. 4] the detent, instead of pivoting and being returned by a spring, was formed at the end of a thin spring. The balance deflected the detent spring to release the tooth and the spring itself returned the detent to lock the wheel. This, and the final development of the temperature compensation so that the bimetallic strips formed the rims of the balance, substantially improved the constancy of rate.

John Arnold died in 1799. His simple and ingenious machine continued to be produced in great numbers by his son, John Roger. Its accuracy enabled the Royal Navy, upon which England depended for her survival, to make a landfall anywhere in the world with complete certainty. Its contribution to England's supremacy in horology and the accumulating wealth of the nation at that time cannot be overestimated.

The final development of the timekeeper to the form in which basically, even today, it is manufactured by Messrs. Thomas Mercer of St. Albans, is to the credit of Thomas Earnshaw. At the age of thirty-one he had evolved an escapement [Fig. 5] similar in principle to Arnold's in that it also employed a spring detent and the sequence of events was the same. The one important difference lay in the position of the detent which, in Earnshaw's escapement, was diametrically opposite that part of the 'scape wheel acted upon by Arnold's detent. This necessitated the detent being in compression when locking the wheel and, although this would, at first sight, seem to be a disadvantage, it was in fact quite satisfactory. A slight alteration to the proportions of the impulse pallet and improved techniques in manufacturing the balance resulted in timekeeping superior even to Arnold's. By his work on the chronometer Earnshaw showed himself to be a most intelligent mechanic whose work was an important factor in the development of the precision timepiece. He made a large number of pocket watches employing his escapement and firmly established its reputation. During his stormy lifetime his opinions, couched in most forceful phrases, led him into many clashes with the Board of Longitude, and also with John Arnold, whom he vehemently, and possibly truthfully, accused of stealing his spring detent escapement. He eventually received a reward of £2,500 for his contribution to the timekeeper, but hotly insisted that this was insufficient. Since his spring detent was basically an extension of Arnold's pivoted detent and he made a great deal of money from its manufacture he really had very little to complain about.[1]

The watches of Arnold and Earnshaw have usually plain single cases in silver or gold. The winding is accomplished by opening the back and inserting the key into a hole in the inner non-opening dome of the case. Dials were plain enamel with, in later examples, the elimination of the archaic minute numerals, as it seemed reasonable to expect the wearer to work out the exact time without their assistance.

[1]For further details, see *The Marine Chronometer*, by Lieutenant-Commander Rupert T. Gould.

6. *The Early Lever and Duplex Escapements*

ALTHOUGH THE MERITS OF MUDGE'S LEVER ESCAPEMENT WERE GENErally unrecognized during the period of development of the chronometer it still had its champions. Chief among these were Count von Bruhl, diplomatic representative of the Kingdom of Saxony to the Royal Court of Great Britain, who had been Mudge's patron, and Mudge's son Thomas.

Count von Bruhl prevailed upon Josiah Emery, a Swiss born watchmaker resident in London, to undertake the construction of some watches employing Mudge's lever escapement. Although reluctant to start the work and complaining that the task was too difficult, he nevertheless produced the first watch in 1782 and included in its specification the compensation balance devised by Arnold in 1779. Constructed to better proportions than those adopted by Mudge it was much more vigorous in its action. This, combined with the superior compensation in the balance, assured it a greater success.

Its timekeeping did not equal that of the chronometer but, in wear, it proved to be more reliable. The chronometer, receiving as it does its impulse at each alternate vibration is, if roughly handled, inclined to stop. The lever, with the vigorous action occasioned by the lighter balance and larger impulse radius as constructed by Emery, was, with its impulse at each vibration, less affected by rough usage. Undoubtedly such a watch was more difficult to make than the simpler chronometer. The pallets required shaping entirely by hand and this was a tedious and time-consuming operation.

Emery continued to make these watches until his death in 1797, producing some thirty-six pieces altogether. He made various alterations to the escapement, none of which improved its performance but merely simplified its manufacture.

He seems, however, to have given little thought or attention to the prime fault in the escapement. This fault can occur during the supplementary arc when, after the lever has completed the impulse, it rests against the banking pin waiting for the balance to return and re-engage the fork. If at this stage the watch is moved sufficiently forcefully to dislodge the lever, the safety dart rests against the slotted roller. This action prevents the lever accidentally unlocking and so preserves the working of the watch, but at the same time the interference with the balance robs it of its timekeeping.

The arrangement of the escapement as used by Emery is known as the 'straight line lever' because the pivot holes of the balance, lever and 'scape wheel are planted in a straight line. As designed by Mudge in his watch it is known as the right angle lever – the holes being arranged in a triangle with the lever to one side of the 'scape and balance holes. Considering that one of Emery's objections to making the escapement was the difficulty of compressing it into the space available in a watch it is curious that he should have chosen this layout. It may be that he anticipated the subsequent charge by Thomas Mudge junior that any success Emery had had with his watches was attributable entirely to his father's escapement. By adopting the straight line plan Emery may have intended to show that he was not blindly copying but quite capable of making an individual approach. Whatever the reasons Emery's watches are the first straight line development of the detached lever.

Others making watches along similar lines were Grant, Pendleton, who is credited by Mudge's son with the construction of the Emery watches, Perigal and Dutton whose watches also have Pendleton's style. These beautiful and complex watches all suffer from the same fault but nevertheless give a reliable approximation to time.

One maker only saw the necessity for overcoming this fault, that is, the friction caused by the safety dart and he, John Leroux, solved the difficulty by introducing a small angulation of the locking faces of the pallets. Thus when the lever was dislodged from its banking it caused a recoil in the 'scape wheel. The resistance of the 'scape wheel to the recoil returned the lever smartly to its banking and so relieved the balance of the friction of the safety dart. With Leroux's improvement and Grant's construction the lever could have been universally used in pocket timekeeping. Alas! In the hands of Emery and others it had earned for itself an indifferent reputation and this, coupled to the difficulty of its construction, ruined its chance of success.

These difficulties could of course have been overcome, as indeed they were later, but the introduction in England in 1782 of the duplex escapement [Fig. 3] offered a brighter prospect of cheaper precision watches. The escapement, of French origin, was patented in England by Thomas Tyrer, of whom nothing further of note was heard. It, like the chronometer, delivers its impulse to a pallet on the balance arbor, the teeth for this function being vertical. After the impulse is delivered long radial teeth on the same wheel are arrested by the balance arbor whose radius they intersect. As the balance approaches the next impulse vibration, the long tooth, resting against the staff, passes through a slot cut in the balance arbor, and the wheel, in rotating until the next long tooth takes up the locked position, delivers an impulse to the balance with the vertical teeth.

In the country of its origin it met with little success, for the French favoured thin watches and the escapement was not suitable for these. The English liked thick watches, or meekly took what they were given, and, constructed as it could then be, it was an immediate success.

In the hands of the best makers of the day and combined with the compensation balance, its performance was, by 1800, superior to the lever. Easier to construct than the lever, it found great favour with the London makers who became very adept at determining the best proportions. So skilful in the construction of the escapement were they that, with its greater robustness and consequent reliability, it became, for pocket use, a serious contender for the honoured position of esteem held by the chronometer. So great were its merits that it was not until the middle of the nineteenth century, when watches, under pressure of the influence of Breguet in France, had become noticeably thinner, that it eventually lost favour. However, it was still occasionally made with the new and thinner three-quarter plate movement. In this arrangement the balance was sunk below the level of the back plate, of which a quarter was cut away to provide room for the escapement. The 'scape wheel pinion was reduced in length by two thirds to fit into the confined space, and these conditions were not suitable to maintain the reliability. Although the escapement had maintained the prosperity of the trade for fifty years it had outlived its usefulness and could contribute nothing more to the future of the industry. The lure of easier profits for its makers had undoubtedly, for a quarter of a century, frustrated the development of the more enduring lever, which by 1815 had begun to reappear in what was to be, for the English industry, its final form.

7. The English Watch: Final Development and Decline

THE LAST PHASE OF THE DEVELOPMENT OF THE ENGLISH WATCH BEGAN in Liverpool in 1792 when Peter Litherland patented his rack lever escapement. Similar in principle, but improved in detail, it owed its origins to the escapement invented by the Abbé d'Hautefeuille in 1722.

The manufacture of special jigs and tools for producing the pallets, and the use of the rack and pinion for imparting the impulse to the balance, almost eliminated hand work in the construction of the escapement. These methods ensured its success and allowed many thousands to be produced on a semi-mass production basis. Many Liverpool makers turned their attention to its manufacture and an export trade to America was started where they found great favour for their reliability and good timekeeping. Superior in this respect to the cylinder, and, by Liverpool methods, cheaper to produce, the rack lever commanded the respect of its down-to-earth Midlands purchaser. The London makers continued with the duplex and chronometer for high grade watches, using the cylinder for cheaper products. They regarded the Liverpool product with contempt and condemned it for its allegedly poor quality.

In 1815 Edward Massey, a Liverpool maker, introduced the first detached variation of the escapement. This he achieved by the simple and ingenious method of removing from the rack all but two of the teeth, these two forming the fork of the conventional lever. From the balance pinion all but one of the leaves were removed and this was shortened by removing the ogival tip. At each side of the root of the leaf a slot cut into the axis of the pinion allowed the single leaf pinion and forked lever to engage more deeply. Accidental unlocking was thus prevented by the fork of the lever rubbing against the roller. During the impulse the tips of the fork passed into the slots of the roller which ensured their free passage.

The resulting timekeeping was an improvement on the rack lever where the interference of the rack with the pinion is a constant distraction to the rate. In later examples the pinion leaf is replaced by a jewel set in the steel roller and, still later, it became a jewelled pin set vertically in the roller.

By 1820 the London makers began to take an interest in the escapement. For the previous five years Breguet had been producing extremely thin and elegantly proportioned watches. Their lever escapements of most advanced design were entirely the result of his own development. With their increasing appearance in England the English watchmaker was forced to realize that he was in danger of losing his reputation as the constructor of the world's finest watches.

In about 1820 George Savage introduced his 'two pin' lever escapement in which the vertical pin of the later Massey lever was replaced by two brass pins set apart at the edge of a flat roller. These two pins served to perform the unlocking. The impulse was delivered by a third pin mounted in the end of the lever, at the centre of the fork, and engaging a slot in the edge of the roller. After completing the impulse the pin passed out of the slot and served as a guard pin to prevent accidental unlocking.

By a combination of the single pin of Massey for the impulse, and the pin and slot of Savage for the safety action, the final solution to impulsing was reached. There remained one point in which the English makers were ahead of their rivals and this was in recognizing the necessity for keeping the balance free of interference by the lever after the impulse was completed. This was achieved with the technique, introduced by Leroux in his earlier escapement, of giving a positive angle of locking to the pallet stones, causing the recoil of the 'scape wheel to draw the lever back to its stop, or banking, if displaced.

Even the illustrious Breguet was reluctant fully to recognize this necessity which, by 1820, was completely acceptable to English makers.

Once again English watches were supreme in their simplicity of construction and timekeeping. The rate of the lever in this form [Fig. 6] was so close as to rival the chronometer for pocket use. By the middle of the nineteenth century the chronometer began increasingly to give way to the lever which also, being more robust, was better suited to pocket use.

For the remainder of the nineteenth century the lever almost completely dominated the trade, although the chronometer retained its position as the supreme timekeeper, with the duplex finally abandoned by the end of the century. The indestructible verge fell into disuse in general commercial production only a little before this but there is some evidence that it was still being made to individual order in the twentieth century.

Dials were almost always white enamel with thin elegant Roman numerals, although some Earnshaw dials had Arabic numerals. Cases of silver or gold became flatter. In later examples both back and inner dome could be opened, thus it was no longer necessary to hinge the movement out of the case in order to examine or regulate it. Barleycorn engine turning became increasingly popular but began to wane by the third quarter of the nineteenth century. With this form of decoration the effect is achieved by oscillating a revolving mandril, carrying the work, against a vee-shaped cutting tool.

The keyless work, introduced by John Roger Arnold after the design of his workman Thomas Prest, was a crude beginning to the eventual abolition of the winding key. In the early Prest winding the hands are set by pushing them with the finger. Later development resulted in the provision of a pin or lever that, when moved, engaged the hands with the winding crown.

In lower grade watches the fusee was dispensed with and stop work used to make a compromise with the spring torque. In this arrangement, called going barrel, the great wheel is fixed to the barrel and rotates with it during the unwinding of the spring.

To secure the advantage of Arnold's helical spring curves without the disadvantage of the increased height of the spring, Breguet's form of spring became more frequently used. In this the last coil of a spiral spring is curved up and in to a smaller radius. Thus the flatness was achieved and the advantage of the manipulation for escapement error retained.

It is the greatest tragedy of the English watch industry that, having overcome, in two hundred and fifty years of development, every obstacle in its path, it then allowed itself, through prejudice and lack of enterprise, to go into a decline. The revolution that was taking place in social behaviour and customs opened up a vast new market for a cheaper watch. With studied conservatism the trade, apart from some abortive and ill supported attempts at mass production, refused to supply the market. The mixture of contempt and fear with which they viewed the changing conditions reduced the industry to a shadow of its former importance. The few remaining masters of the art such as J. F. Cole, C. Frodsham, E. J. Dent and Viner continued to make, for an exclusive market, the most individual watches and carriage clocks. The remainder continued to produce hand finished levers and chronometers which only a small percentage of the population could afford.

The Swiss, who were not hamstrung by tradition and prestige, had equipped their industry to mass produce the best inventions of the English artists, and enterprising merchants soon began importing them in to England. Employing mainly the cylinder escapement and cased in the English style the watches proved popular and reliable. Increased importation, at prices the English makers could not compete with, still further reduced the number of English watches produced.

As a last defiant gesture the English maker showed that if he didn't intend to submerge his identity in the factory system he was still the master of timekeeping. The watches constructed by him, and containing the complicated and delicate tourbillon and karrusel rotating escapements for the elimination of positional errors, could maintain their rate to a variation of less than a second a day.

By a combination of climate and the seasonal circumstances of industry associated with it, the Swiss prospered in their expanding industry. By the application of clever scientific horology their products were continually improved. The introduction during the 1914-18 war of small precision lever watches specially designed for wrist wear emphasized the difficulty of making a fresh start in England that had any hope of succeeding.

The only development of note that occurred in England between the end of the First World War and the beginning of the Second was the invention in 1922, by John Harwood, of a mechanism for automatically winding a wrist watch while in wear. Harwood, a watchmaker of Bolton, Lancashire, had for some years, before taking out his patent of 1924, considered the possibility and the advantages of automatic winding. These advantages are the improved maintenance of the balance amplitude, through the continuously wound state of the mainspring, and the exclusion of dirt by dispensing with the winding stem.

The weight, which was to be oscillated by the movement of the wrist and so wind the spring, was, in the early experiments, pivoted near the edge of the movement after the manner of Breguet. This arrangement proved unsatisfactory and the weight was then pivoted at the centre of the movement. This position is now accepted by many manufacturers as the most satisfactory. The hands were set by rotating the knurled bezel and so the stem was eliminated.

Harwood took his watch to Switzerland with the hope of persuading the Swiss manufacturers to produce it in quantity. The Swiss, with a conservatism that rivalled that of the early English makers, saw no future for the invention, and Harwood returned disappointed to England.

Back in England he had the good fortune to secure enough capital to cover the cost of tools and labour for a new effort to start manufacture, and, thus ensured against financial loss, the firm of A. Schild S.A. of Switzerland undertook the manufacture of the watch. Its production entailed many problems which Harwood's application and energy eventually overcame. This effort culminated in his personal assembly of the first three dozen movements.

With success in the venture seeming certain, factories were established in France and America, with a head office in London. It was a cruel blow to a spirited and determined enterprise when the world trade slump of 1931 caused the company to cease activities.

After the Second World War Messrs. Smith's of England started to produce wrist and pocket watches comparable in quality to the Swiss product in the same price category. Tremendous difficulties have been overcome and the venture has proved a success. If the industry can emulate the inventiveness and ingenuity of its forbears it must continue to thrive. If it ever rests, as they did, upon its laurels, it will surely fail again.

8. *American Watchmaking*

THE EARLIEST WATCHES OF AMERICAN CONSTRUCTION WERE NECESsarily of European style and finish. Made by immigrant craftsmen from England, who were joined later by watchmakers from Holland and Switzerland, their work and methods followed the dictates of their training. With no established industry to facilitate division of labour, a watchmaker would, in many cases, be obliged to manufacture the whole of the watch himself. The cases and dials would present a great problem, for although it might have been possible, from the earliest days, to import such items the delays and difficulties would be considerable. Not the least of the craftsman's difficulties would be the problem of finding enough clients able to purchase his products, for the sparse population found life difficult enough without the added burden of purchasing watches. Clocks were a somewhat different proposition, for even before 1800 it was not easy to run a house without some means of telling the time. The making of clocks was less exacting than watchmaking, and the refinement of finish, so necessary in a watch, was not so essential to the going of a clock, even though a better rate could be achieved. Like the watches, they owed their style and construction entirely to European influence. The gradual growth of the industry allowed the training of apprentices and some of the more successful makers were able to employ a few skilled men.

The tenacity with which these craftsmen pursued their difficult task accords well with the spirit of the early pioneers. The increased importation of rough materials from England considerably increased the number of watches produced and lowered the prices, with consequent expansion of the market.

By 1809 Luther Goddard had established the first watch factory in America. It is probable that the movements were based upon English *ébauches* or rough movements. Even so the impressive total of five hundred finished watches was sold before he retired in 1817.

In 1838 an attempt to mass produce the watch was made by James and Henry Pitkin at Hartford, Connecticut. This ambitious project produced some eight hundred watches, with some parts made with power driven machinery specially designed for the work. Inadequate administration and failure to find a firm market ended the venture in 1841. Although in itself a failure, the enterprise of these two men had shown that the manufacture of the all-American watch was a feasibility. It was possible, without recourse to imported parts, to make the machines, find enough skilled men to operate them, and assemble the components of a watch designed in America. The next decade was to see the beginning of the great American factories which were eventually to succeed in doing this.

The first self-contained factory was started in Boston, Massachusetts, by Aaron Dennison and Edward Howard. Dennison, born in 1812 at Freeport, Maine, had been apprenticed to James Carey, a watchmaker in Brunswick. After completing his apprenticeship he moved to Boston where he was employed in the repair of watches and learned to appreciate fine quality work.

Later he started a business repairing watches and dealing in fancy goods and materials for the trade. His enterprise enabled the business to prosper and it was not long before his thoughts turned to the prospect of producing his own watches. He had had no experience in manufacturing watches, but he was a man of vision and, as is so often the case, the more he thought about the subject the more grandiose became his ideas. His ideal was to be a factory in which the watch was hardly to be touched by hand. All the many intricate machining processes were to be done by automatic machines. Once started they would turn out a succession of components; only resetting the tools would interrupt the process. The machines would make components to precise limits, each one identical and interchangeable. This was a vision indeed, and one worthy of much effort to accomplish.

It was in Boston that Dennison met Howard. Born at Higham, Massachusetts, in 1813, Howard was apprenticed to the son of the famous American clockmaker Aaron Willard. He showed himself to be an intelligent and resourceful mechanic.

At the age of twenty-seven he started in business himself and two years later, in partnership with D. P. Davis, he started a small factory for the manufacture of domestic and precision clocks. His interest, continually distracted by thoughts of watches, turned to the idea of a factory for their production. Like Dennison he visualized quantity production of precision watches by automatic machinery. It is doubtful if any other horologists in America had at that time the ideas and confidence of these two men.

When they had formulated their ideas and worked out the plan to be adopted they began seeking the capital to found their enterprise. This was difficult to raise, for American watches were not popular, English makes being preferred. Samuel Curtis of Boston was eventually persuaded to put twenty thousand dollars into the enterprise, to be known as the American Horologe Company.

The first watches were produced in 1853 under the new name of the Warren Manufacturing Company. They sold at forty dollars. These watches were constructed in the English full plate style. In 1854 the company moved to Waltham where production was started again. By now, under the name of the Boston Watch Company, some ninety men were employed producing thirty watches a week. Howard's insistence on making all the components in the factory was beginning to undermine its stability. As much time was spent in overcoming production difficulties as in actually manufacturing parts.

By 1856 things were getting desperate. Production experiments and the building of machinery took up a great deal of time and money. In 1857 the company was bought by Tracey & Baker, case manufacturers. Dennison stayed with the reorganized company as works manager. Howard, who was made of sterner stuff, returned to his factory at Roxbury to make a fresh attempt at manufacturing a factory produced watch. This he did by restarting the Boston Watch Company. In 1861 the company was expanded and named the Howard Clock & Watch Company. In 1881 it was renamed the E. Howard Watch & Clock Company. A year later Howard retired. He died in 1904, having realized his ambition to produce and establish the American precision watch. To him surely must go the honour of the designation, 'father of American watchmaking'.

Dennison, after disagreeing with the policy of the company, resigned and came to England in 1870, where he founded the case company which still bears his name.

Tracey Baker & Company was eventually, after many changes, to become the Waltham Watch Company.

The earliest watches produced by the Howard Watch Company were fitted with uncompensated balances and flat spiral balance springs. A three-quarter plate model soon followed and this had the compensated balance and a patent mainspring barrel. In this the mainspring was fitted into a stationary barrel and wound by rotating the arbor anti-clockwise. The power, transmitted via the winding ratchet upon releasing the key, was maintained during winding by a form of Harrison's maintaining power. The object was to prevent damage to the wheel teeth in the event of the mainspring breaking. The escapement was lever, with divided lift and single roller. The count of the train was 18,000 and the design was very advanced by any standards of the time.

A later and equally successful model was keyless. This also employed a safety barrel of somewhat different pattern to accommodate the design of the keyless work. These were originally fitted with J. F. Cole's resilient escapement in which the shock of overbanking is absorbed by recoiling the 'scape wheel. The marketing agents claimed that they were unsatisfactory and they were later converted to the conventional wheel and had banking pins fitted. The movements are high quality and fully jewelled in the English style.

Howard continually experimented with improvements for his watches and never relaxed his standards.

Many watch factories were started in the last quarter of the nineteenth century and first quarter of the twentieth century. Most of them failed within five years of starting. Companies amalgamated with each other in a complex and interwoven pattern under an endless variety of new company names. That the average American had accepted the American made watch there was no doubt: he could hardly be blamed however for preferring to purchase the Swiss equivalent at a lower price.

There is one precision watch, quantity produced, that the American industry can be particularly proud of. This is the presentation Waltham Riverside Maximus. This superb watch, always in a heavy gold case, is worthy of a place in any collection. Fully jewelled and with diamond end-stones to the escapement, it is beautifully finished in machined nickel. The keyless winding mechanism operates with a smoothness rarely equalled by any other maker and the general attention to detailed finish is worthy of the highest praise.

At the other end of the scale come the Auburndale, Waterbury and Ingersoll watches. The Auburndale Rotary watch was invented by Jason Hopkins, a manufacturer of pianos and clock cases, and patented by him in 1875. He took his prototype to Washington where he was introduced by his patents attorney, W. D. Colt, to the industrialist Edward Locke, who displayed great interest in its possibilities for mass production. Edward Locke took the watch to Boston where he persuaded George Merrit, a financier, to put up sufficient money to improve the prototype and better its chances of success.

The watch was put into production in 1875 but the factory soon found itself in difficulties through insufficient sales. In an attempt to improve the financial position the company was reorganized as a public company in 1879. In the same year a patent was taken out for a stop watch. This proved a success and was followed by a split seconds timer. During the horse racing season this sold well but out of season sales dropped alarmingly.

In 1881, to help keep the factory going, a pocket thermometer was introduced to the catalogue and described as the Auburndale metal thermometer. By 1883 the company foundered through insufficient sales in a depressed market.

The Waterbury, employing a cheap version of the duplex escapement, was patented in 1878 by D. Buck of Worcester, Massachusetts, who constructed the prototype himself. It was first manufactured at the Waterbury factory of the Benedict & Burnham Manufacturing Company and thus it received its name. Soon, requiring more space for production, a new factory was built and the Waterbury Watch Company was formed.

The first watches appeared in 1880 and were soon successful. Eight years later the company was producing half a million watches a year. The ingenious movement was simple to produce and the complete watch cost four dollars. The only adjustment required was regulation for mean time. The whole movement rotated once an hour and carried the minute hand. The mainspring, housed in the back of the case, was some nine feet long and took an inordinate time to wind. This irritating feature, largely responsible for its wane in popularity, currently earned for Waterbury the title of 'Land of everlasting spring'.

The number of types manufactured runs through the alphabet from A to W, with apparently no omissions. The differences in each model are small, and all models up to Series E had rotating movements. In addition to this model a new series was started with the more conventional arrangement. This new model eliminated the long winding necessary with the rotating movements. The earliest Model A had a flat sided winding crown and paper dial. From Series B the mainspring cap advised the tyro that it was to be removed only by the experienced watchmaker! The dial of these and of all subsequent models was unskeletonized although some Model E watches reverted to this earlier practice. From Series K the balance was under a separate cock and the dial sometimes had a sunken centre. This series carried the name Charles Benedict instead of the Waterbury trade mark. Series R assumed great sophistication and displayed an enamel dial with subsidiary seconds. These and later models could be bought with silver or rolled gold cases. They came in a silk lined box complete with a thirty-two page booklet. At one stage in its existence the company were making one thousand watches a day.

After some ten years of production the company abandoned this series, and under the name of the New England Watch Company introduced a new model. Failing in this venture, they were then taken over by Ingersoll.

R. H. Ingersoll and his brother C. H. Ingersoll had started to sell cheap watches in their mail order business in 1892. The watches, manufactured by the Waterbury Company, sold for one dollar only. They were robust to the point of crudity. A vigorous action induced by a very fierce mainspring ensured their going, although the rate could not always be relied upon.

This company failed in 1922. During its period of activity it produced over sixty million watches. By supplying reasonably adequate time to anyone who could find a dollar, the watch itself created a demand for a better product. The American factories were, on the whole, unprepared for this new market which was eventually dominated by the Swiss wrist watch. The manufacturers of the cheap American watches had neither sufficient experience to make a better quality product nor the reputation needed to sell them. The high quality products of Waltham, Elgin, Illinois and Hamilton could not compete with the Swiss system of division of manufacture. Broadly speaking, this system entails the delegation of the manufacture of components to specialist manufacturers. With increasing experience in their own field manufacturers were able continually to improve quality without increasing the price of the completed watch.

During the Second World War production of the American factories was increased to capacity to supply the demands of the services, the Hamilton Company earning particular merit with the mass production of their superb marine chronometer. The rapidity with which the machine was developed and the ingenious form of compensation employed illustrate the ability of the industry. The production of ten thousand such machines in three years is a most remarkable achievement and one which could have occurred only in America.

In the production of the mass produced watch, factory tool rooms showed the greatest ingenuity in the construction of the necessary machinery. Machines were devised and constructed by the factory staff which could perform all of the operations necessary to the machining of the component, and then impart the final polish. A blank pressing, automatically fed into the machine, would emerge as a finished plate after receiving turning, sinking, drilling and tapping operations. Regulator indexes and keyless parts were stamped out and fed into automatic lapping machines that followed every change of contour and section, delivering the piece polished to perfection. Jewel holes, screws, pinions and wheels were all made in the same manner. Today there is not a watch factory in the world that does not owe something of its success to American inspired methods.

The individual watchmaker producing handmade or hand-finished precision watches is rare in American history. Of the select few who accomplished this feat some made only a few, rather more to satisfy their own desire than for any financial reward. Two are particularly noteworthy as having devoted their lives to the task of making their own individual watches to the highest standards.

Charles Fasoldt, born in Dresden, Germany, set up as a watchmaker in New York in 1850. In 1861 he moved to Albany, N.Y., and started a factory where he eventually employed fifty or more men. High quality regulator clocks and pocket watches employing Fasoldt's patent escapement were made. The escapement had two wheels mounted on the 'scape pinion. The larger of these performed the locking and unlocking of the 'scape wheel in conjunction with the conventional lever pallets. The smaller wheel delivered the impulse, at each alternate vibration, to a pallet fitted to the lever arbor. The intention was to eliminate oil at the escapement and also secure the advantage of the lighter locking afforded by the larger locking wheel. The escapement works well although no better than the conventional lever.

This very individual, if somewhat elaborate, escapement was, with some alteration, later used in his regulator clocks but there is no record of their performance. The watches are most beautifully finished, all the steel work is polished and the jewels are set in gold chatons. Regulation was by Fasoldt's patent micrometer regulator without backlash. Most of his watches show individual differences of detail finish, depending on the whim of the finisher. Some had a constant force spring between the fourth wheel and pinion to equalize small variations in power. Since the tension in the spring is directly proportional to the power of the mainspring it achieved nothing in a day's run.

Albert Potter, born in New York, was apprenticed in 1852, and started business on his own account in 1856, still in New York. Between repairing watches he made, entirely with his own hands, some thirty-five watches employing the lever and chronometer escapement. Moving his business from place to place, he continued with his work, producing another twenty or so watches, and including, in some, his own repeating work. In Chicago, with his brother William, he started a business called Potter Bros. in which he continued to make individual and beautiful watches. In 1876 he settled in Geneva, Switzerland, and began seriously to make the watches for which he was to become famous. In his small establishment he made lever and chronometer watches to his own design. His escapements were beautifully proportioned and finished. The movements were finished in machined nickel and the general standard of finish bore favourable comparison with the best work of the English makers.

At Charmilles a factory was built to manufacture a cheap, good quality watch to Potter's design. The establishment at first showed signs of succeeding but eventually, in 1896, failed for lack of public response.

His handmade watches were so expensive as to interest only a small and discerning market. Many fine watchmakers have found, as Potter did, that the existence of the true artist is a precarious one. Disappointed in his endeavour, he partially turned his interest to other things. At the age of sixty-five he was suffering from paralysis resulting from a spinal disease, and he died in 1908 a disillusioned man.

9. *Post-War Developments in America*

THE IMMEDIATE POST-WAR YEARS SAW LITTLE CHANGE IN THE DEvelopment of the watch in America. The factories, more firmly established after their wartime production runs, continued to produce pocket and wrist watches for home and export markets.

The fact that the Swiss factories were making continued progress with automatically wound watches seems to have caused the American industry not the slightest concern. The automatically wound watch, being continually wound by the movement of the wrist, was better equipped to maintain a smooth rate. Isochronal errors that would occur with the conventionally wound watch were much reduced by the mainspring being kept in a fully wound condition and thus maintaining the balance amplitude.

The invention of the automatic watch is attributed to Abram Louis Perrelet, a talented Swiss watchmaker, in 1775. It was further developed and manufactured by Breguet but attracted little attention after his death in 1823. It was not until 1924 that interest was re-awakened by the English watchmaker, John Harwood, who patented and produced a mechanism for automatically winding a wrist watch in wear. The venture was brought to a financially unsuccessful conclusion by the trade slump of 1931.

It was to be presumed that the American industry would consolidate its position by introducing some new models to challenge the lead of the Swiss, but it was not until 1957 that the Hamilton Company announced their answer to the automatic. This, an electrically powered watch, was quite revolutionary.

The movement, although simple in appearance, no doubt presented its makers with some considerable technical problems. The performance and reliability of the watch have shown them to be most successfully overcome.

The balance wheel is impulsed electro-magnetically, a coil of wire attached to the balance being attracted to a permanent magnet when energized by the battery at alternate vibrations. The contacts for completing the circuit to the coil are fixed to the plate of the watch and closed, in passing, by a pin on the balance staff. The hands are advanced by a suitable train of gears commencing with a star wheel which is advanced one tooth per oscillation of the balance by a second pin on the staff.

It might be thought that, in receiving its impulse at alternate vibrations, the watch would be likely to set, as can happen to the chronometer escapement. This would in fact never occur in normal use, for the balance needs to be turned through only a very small arc to close the contacts and receive an impulse.

The advantages of this system of timekeeping are, elimination of wear, since the train transmits no power, and considerable reduction in isochronal errors which arise out of the variation of impulse received by the conventional watch balance.

A more recent advent on the American horological scene has been the Bulova Accutron electronic watch.

The Bulova Watch Company of America was founded in 1875 by Joseph Bulova. In its beginnings it was simply a shop for making and selling jewellery but it was the ambition of Bulova and his son Arde to manufacture fine watches. The company, very successful in its aims, now has factories in Germany, Switzerland and Canada in addition to its American factories. Of the range of some 600 different types and styles of watch manufactured by the group of companies under the name of Bulova the most remarkable is the Accutron.

The movement of this watch is a complete departure from the conventional electric watch. In its elements it consists of a small vibrating fork carrying a permanent magnet at the tip of each tine, two small coils to impulse the magnets, and an electronic circuit comprising a battery, a transistor and a capacitor with its attendant resistor. One of the impulse coils is divided into two parts. The larger part impulses the tine magnet and the smaller part induces a current into the capacitor. With the fork vibrating, the permanent magnet enters the coil and generates a current in the smaller coil. The current is fed to the capacitor which, when the designed value is reached, discharges into the transistor. The resultant charge of current to the transistor causes it momentarily to become a conductor completing the circuit between the battery and the impulse coils. The tines of the fork, in returning to their natural position, again cause the small coil to charge the capacitor and the process is repeated with the forces in opposite polarity.

The vibrations of the fork are converted to rotary movement by an indexing jewel attached to one of the tines. This engages a ratchet wheel containing 300 teeth in its circumference of only 7.6 mm. From this a suitable train of gears turns the motion work carrying the hands. There being no power transmitted through the train, as in the conventional watch, the pivot pressures are negligible and wear at these parts is elimінted. For this reason maintenance requirements are minimal and the watch will run for many years without mechanical attention.

As a wrist watch requiring almost no attention except battery replacement at intervals of twelve or more months, it sets a very high standard of reliability and consistency in rate. Its accuracy is guaranteed by the makers not to exceed two seconds' variation per day but if the tines are adjusted to suit the individual wearer the error can be reduced to less than half this figure.

The development for production of the watch started in 1952 when Max Hetzel met William Bennet. Both these men worked for the Bulova Company and both had done important work in their own fields before their meeting. Bennet, as a production engineer, had played an important part in the production of watches and chronometers for use by the United States armed forces. Hetzel had made improvements in telephone and telegraphic communications in the field of electronics. In 1954 he made the first wrist watch utilizing the vibrating tuning fork. Its rate of one second per month variation was a most encouraging culmination to several years of development work.

1956 saw the beginning of serious planning between Bennet and Hetzel to produce the electronic watches in quantity at the Bulova plant at Jackson Heights, N.Y., and by 1960 watches were being made in sufficient quantities to enable the company to announce their introduction to world markets.

With such enterprise and inventive ability at its command the American industry should have no difficulty in holding its position as one of the world's leading manufacturers of fine watches.

The direction that the future development of the watch will take is not easy to predict. No doubt it will be increasingly in the fields of electricity or electronics – perhaps even radio. The necessity for maintenance, no matter how little, will encourage technicians to eliminate mechanical working parts from watches altogether. How long this will take depends on a variety of economic and technical problems. That it will ultimately occur seems inevitable. When it does, the watch, excepting perhaps for its appearance, will cease to hold any further interest for collectors and students.

There are, tucked away in obscure places and dark corners of houses all over the world, many interesting and, alas, for the impecunious enthusiast, sometimes valuable watches. Each year a few more come to light and are snapped up by the increasing number of eager collectors. It also seems that many more younger men, their mechanical instincts largely frustrated by the expanding use of electronics, turn to the study of watches for relaxation and interest. Although this makes the acquisition of old watches even more difficult for the collector it is a welcome sign that the work and aspirations of the artists and workmen of the past will always be preserved, and perhaps inspire future generations to emulate their endeavours.

It is this author's sincere hope that the numbers of these enthusiasts will ever increase.

The Verge

THE VERGE ESCAPEMENT IS THE EARLIEST PRACTICAL ESCAPEMENT TO be applied to a watch or clock. Its origins are obscure but its use goes back to the very beginnings of mechanical timekeeping.

It is illustrated in Fig. 1. The pointed toothed 'scape or crown wheel as it is called is carried upon the last pinion of the train. When used in a watch where the balance is required to be horizontal or a clock employing a pendulum the pinion is driven by a contrate wheel to achieve a drive through ninety degrees of angle. In clocks with foliots, resembling a vertical balance without the rim, the crown wheel is horizontal to the plate and so does not require the contrate wheel. Its action is as follows: with the wheel rotating in the direction of the arrow the pallet or flag, forming part of the balance axis, nearest in perspective, will be impelled by the tooth which is in contact beneath. The clockwise rotation of the verge arbor will turn the furthest flag to meet the advancing tooth of the wheel. As the impelling tooth leaves the nearest flag the approaching tooth will fall onto the furthest flag. The momentum of the balance or pendulum will continue to turn the arbor clockwise and thus drive the wheel back. The continued rotation of the arbor is called the supplementary arc. The resultant reversal of the wheel is called the recoil. When the energy of the balance or pendulum is expended its direction must be reversed to allow the next tooth to escape.

With the pendulum the reversal is effected by gravity. The impulse delivered by the 'scape wheel impelling the flags is thus required only to replace the frictional losses in the pendulum, to maintain the clock in motion. The situation is different with the balance wheel for when it reaches the end of its swing the reversal is dependent entirely on the power available at the crown wheel teeth. Thus the balance wheel is at a disadvantage for if the power falls off at the end of the swing it will take longer to reverse and the rate will be slower. Conversely if the power increases it will reverse more quickly and the supplementary arc also will suffer for there will be more power to resist recoiling the crown wheel. Thus the verge escapement is entirely at the mercy of the available power which in early watches was most inconstant.

With the introduction of the balance spring the situation was vastly improved. The balance spring supplied to the balance some of the controlling influence that gravity gave to the pendulum. However, it was still adversely affected by the recoil during the supplementary arc, and therefore, although with the balance spring the verge no longer required the train to reverse its motion, it was still very much at the mercy of its variable influence.

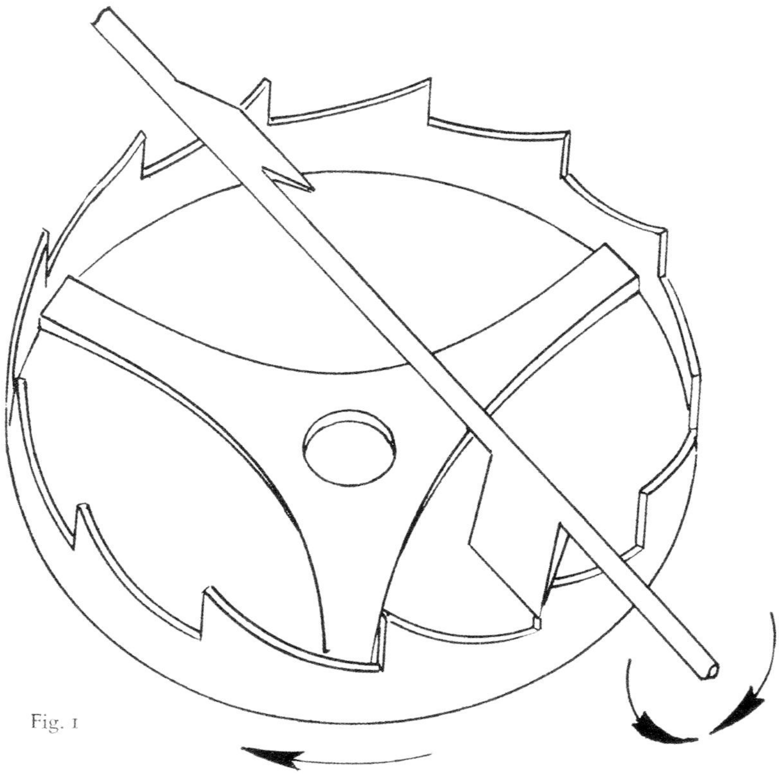

Fig. 1

Cylinder Escapement

THE CYLINDER, SOMETIMES CALLED GRAHAM'S ESCAPEMENT OR THE horizontal escapement because the 'scape wheel is horizontal to the plate, is better disposed to absorb variations in power. It is seen in Fig. 2. The vertical cylinder, much cut away to provide clearance for the 'scape wheel, forms the axis for the balance. The curved inclines, raised above the plane of the 'scape wheel, supply the impulse to the cut away edge of the cylinder which, at this point, is of semi-circular section. The incline or tooth in contact with the edge nearest in perspective is imparting a clockwise impulse to the cylinder. When the end of the incline is reached the impulse or escaping arc is completed. The tooth will drop onto the inner wall of the semi-circular section and rest. The balance continues to rotate clockwise to complete the supplementary arc. When the energy in the balance is dissipated it will return anti-clockwise under the influence of the balance spring. When the point of the resting tooth meets the inner edge of the rotating cylinder it will escape and, in passing, the incline will impart the impulse to the cylinder. The next tooth then falls onto the outside of the cylinder and rests during the supplementary arc. Thus each tooth imparts an impulse first to the outer edge of the cylinder and, on the reverse rotation, to the inner edge.

It is apparent that the cylinder escapement will not function without a balance spring. Unlike the verge there is no recoil of the 'scape wheel to change the direction of the balance. Because there is no recoil the supplementary arc is of a different and superior order to that of the verge. An increase in impulse will still speed the balance but, during the supplementary arc, the increased pressure of the tooth, on the inner or outer surface of the cylinder, will act as a brake and produce the opposite effect. This compensating effect makes the cylinder superior to the verge for timekeeping.

It will be noticed that, for all practical purposes, the radius at which the impulse is delivered equals the resting radius. Also the total arc of the balance is restricted because, although the cylinder is cut back as far as practicable below the acting semi-circular section, it will strike the roots of the teeth if the total arc exceeds some 270°.

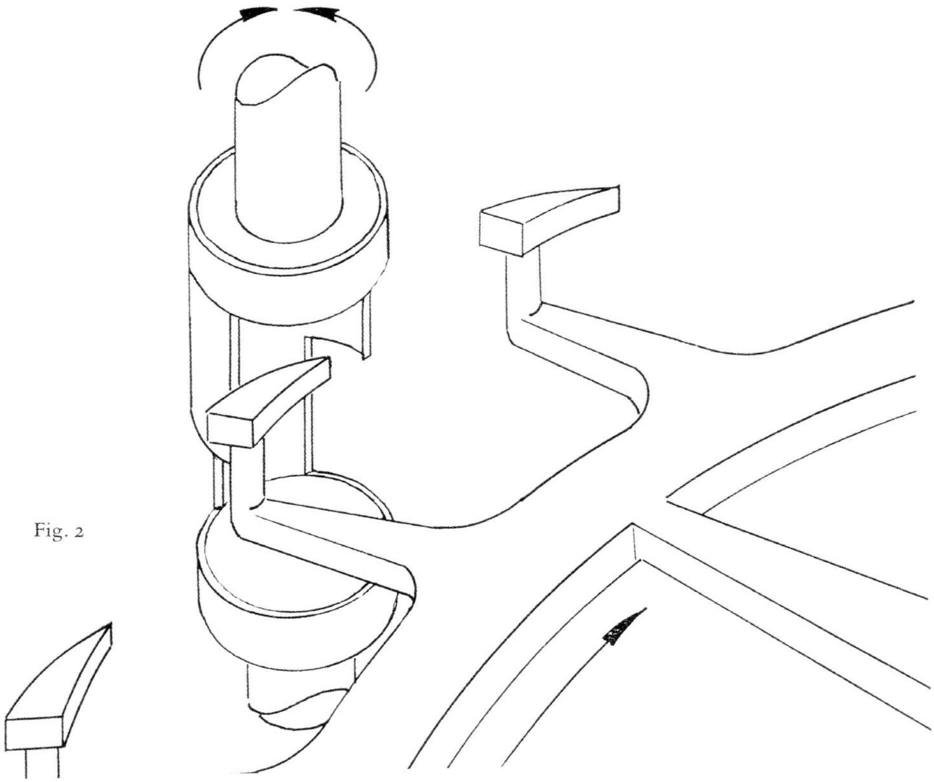

Fig. 2

The Duplex

THE DUPLEX ESCAPEMENT, LIKE THE CYLINDER, RELIES UPON FRICtional rest between impulses. It is, however, superior because the radius of friction is smaller than the impulse radius and there is no restriction of balance arc.

Illustrated in Fig. 3 it can be seen that the 'scape wheel has two sets of teeth, the one radial and the other vertical. The impulse pallet and the cylindrical slotted jewel are fitted to the balance axis. The vertical tooth in contact with the pallet is imparting an anti-clockwise impulse to the balance. At the completion of their intersection the impulse is completed and the following horizontal tooth will drop onto the surface of the cylindrical jewel.

The balance continues to rotate anti-clockwise to complete the supplementary arc. During the return vibration the impulse pallet passes idly between the back of the vertical tooth that imparted the impulse and the face of the following vertical tooth. The horizontal tooth remains resting on the cylindrical jewel during this vibration. The balance, reaching the end of the idle vibration, again reverses for an impulse vibration. The tip of the resting tooth passes through the slot in the jewel as their paths intersect and allows the following vertical tooth to deliver the impulse to the pallet. The resting position is shown inset and illustrates the intersection of the tooth and jewel.

The prime advantages of this escapement are the increased freedom of the balance afforded by the absence of mechanical restriction and the reduced friction of the locking between the small diameter jewel and the long locking tooth. These features are the result of applying the impulse at a separate and larger radius.

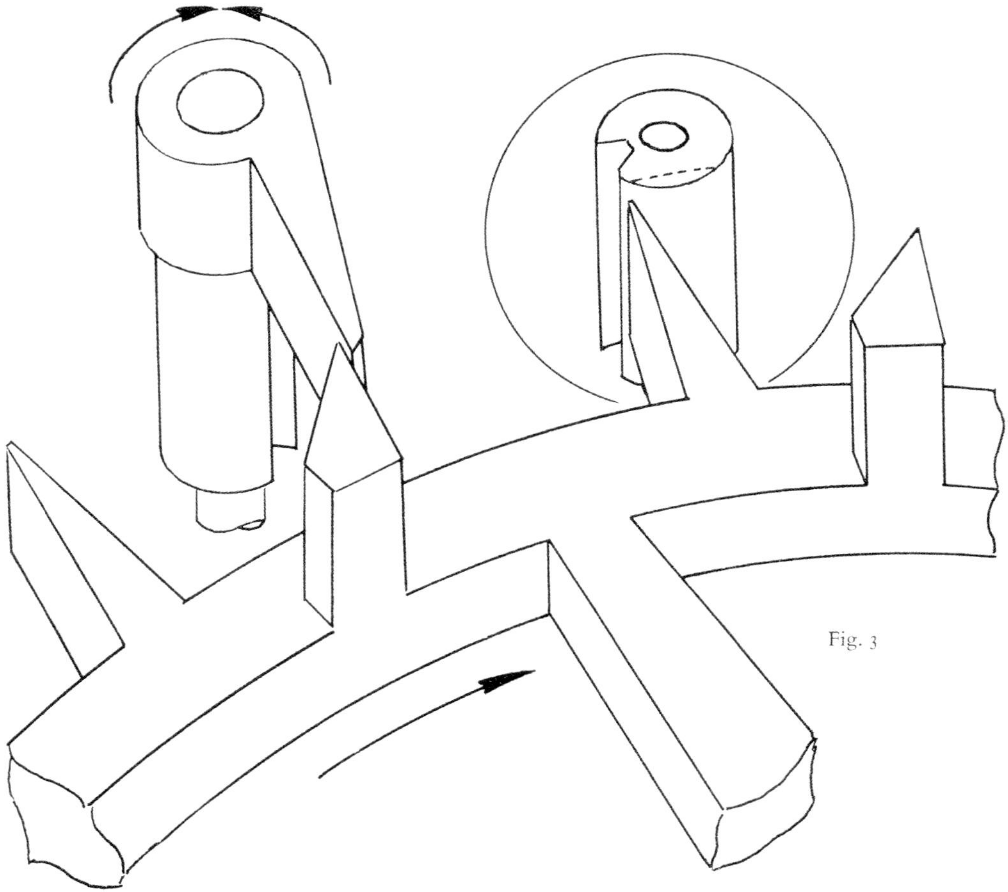

Fig. 3

The Chronometer

THE CHRONOMETER ESCAPEMENTS OF ARNOLD, [FIG. 4], AND EARNshaw, [Fig. 5], can be considered together in so far as their sequence of action is concerned. It can be seen from the illustrations that the prime difference is in the position of the 'scape wheel. With Arnold's the direction of rotation of the wheel keeps the detent in tension. In Earnshaw's the 'scape wheel is moved to the other side of the detent and for the same direction of rotation keeps the detent in compression: Earnshaw's has the single advantage of being easier to construct.

In Fig. 5 it can be seen that there are two rollers carried on the balance axis. Rotating clockwise, the smaller roller has drawn aside the detent by means of the jewel carried radially at its edge. The tooth of the 'scape wheel, in contact with the large impulse jewel, set in the larger roller, is imparting the impulse to the balance. The continued rotation of the balance will free the detent from the unlocking jewel and allow it to return to its original position. At the completion of the intersection of the 'scape wheel tooth with the impulse pallet the impulse is completed. The preceding tooth will fall onto the locking stone of the detent and the wheel will rest. On completing the supplementary arc the balance returns and the unsupported passing spring allows the unlocking jewel to pass without disturbing the detent. Fig. 4 shows this passing action.

In Arnold's escapement the detent is drawn towards the axis of the 'scape wheel during the unlocking. To accomplish this the tips of the teeth of the wheel are raised above the level of the rim and the detent passes over the tips with the locking jewel depending from it. A corner of the tooth is shown locked on the rectangular locking stone. The cyloidal impulse faces of Arnold's wheel are intended to maintain constant the radius of impulse throughout the intersection.

The complete absence of interference with the oscillation of the balance between impulses, and the freedom from restriction limiting the arc of the balance, are the major factors contributing towards this escapement's unsurpassed performance for timekeeping. Unlike the verge and cylinder escapements, the chronometer does not require oil at the teeth of the 'scape wheel and this additional advantage also facilitates the maintenance of its close timekeeping.

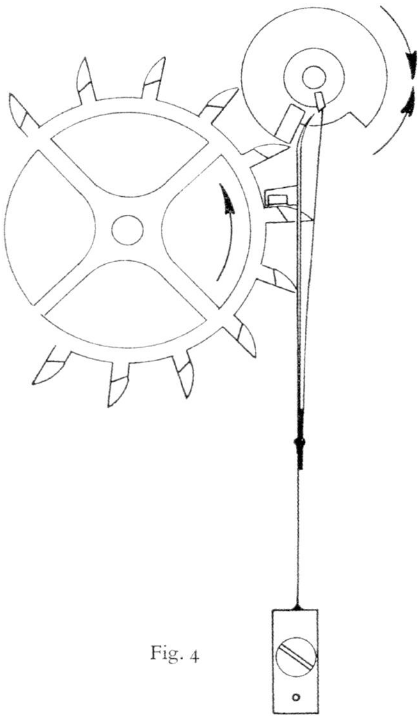

Fig. 4

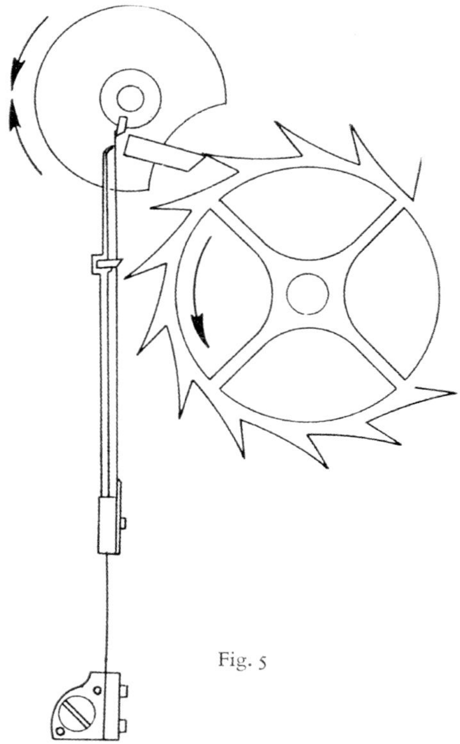

Fig. 5

The Lever

THE LEVER ESCAPEMENT ILLUSTRATED IN ENGLISH RATCHET TOOTH form, [Fig. 6], and club tooth straight line form as devised by Breguet [Fig. 7], is the most satisfactory escapement for personal wear. Like the chronometer it allows complete freedom for the balance after impulsing is completed. Its prime advantage over the chronometer is in the impulse delivered in both vibrations of the balance. Should the balance be brought to rest by excessive agitation of the watch a tooth of the 'scape wheel is always on one or other of the impulse planes. With the balance spring offering least resistance at this point self starting in a well proportioned escapement is assured and the going of the watch maintained.

In Fig. 7 the roller carried on the balance staff is turning in a clockwise direction and the triangular stone will enter the fork of the lever. Its continued rotation will carry the fork of the lever to the left and release the 'scape wheel tooth seen locked on the pallet stone. The tip of the wheel in passing will impart an impulse to the lever via the inclined end of the pallet stone. When the furthermost tip of the incline is reached the impulse is continued by the inclined tip of the tooth. When the tooth leaves the pallet stone the impulse is completed and the wheel is locked by the left hand pallet stone. At the completion of the supplementary arc the balance returns and the sequence is repeated with the balance turning anti-clockwise.

The small roller, allowing the arrow shaped safety dart to pass in the crescent during impulse, was not used in the English escapement. The safety was secured by the vertical pin, at the base of the fork, touching the edge of the impulse roller if the lever was disturbed during the supplementary arc.

It can be seen in Fig. 6 that the impulse is derived entirely from the inclines of the pallets, the stones of which were let in flush with the steel. This is inferior to the club tooth wheel of Breguet in which the clearances can be smaller and so less wasteful of power.

The prime disadvantage of this escapement lies in the necessity for oil at the inclined planes of the pallets. With the deterioration of the oil the rate begins to fall off and for this reason the chronometer is preferred for a consistently close rate.

Fig. 6

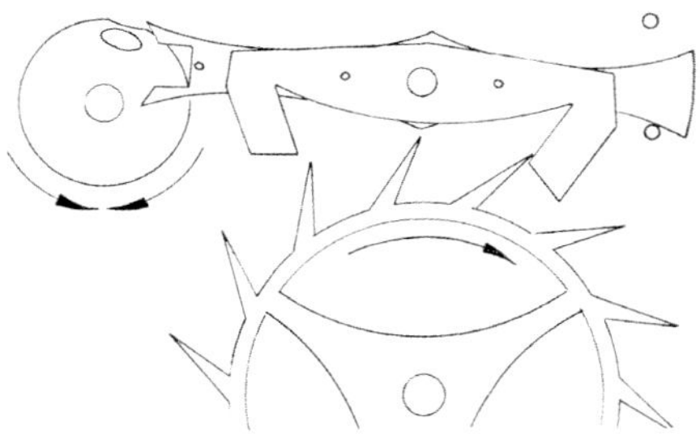

Fig. 7

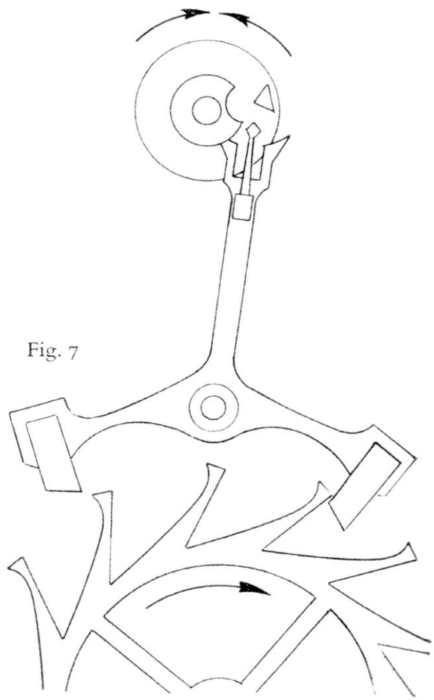

Biographical Notes

OF THE MANY INDIVIDUAL CRAFTSMEN IN THE HISTORY OF WATCH manufacture, relatively few have made a significant contribution to the mechanical development of the watch. Some of the men whose names follow made no positive improvement but because they are, for one reason or another, important enough to be mentioned in the text, a short note of their activities is given here. Those readers who wish to know more about these and the many other important manufacturers who do not appear in this list are referred to the works recommended in the Bibliography.

Arnold, John, 1736-1799. Started business in 1760 as a watchmaker in London. In 1770 he commenced work on marine chronometers and in 1775 patented his compensation balance and helical balance spring. He patented terminal curves to the helical spring in 1776. In 1782 he patented his spring detent chronometer escapement. The business was continued by his son J. R. Arnold.

Barlow, Edward, 1636-1716. Invented repeating work for watches in 1685. He petitioned James II for a patent but was refused on the grounds that Quare's repeating work, invented in the same year, was superior.

Bennet, William O. Vice-President for research and engineering, Bulova Watch Co. of America. Responsible for organization of quantity production of Hamilton marine chronometers and deck watches during Second World War. Co-developer with Max Hetzel of Bulova Accutron electronic watch.

Breguet, Abraham Louis, 1747-1823. Born in Switzerland, he worked for the whole of his life, except for a brief period during the French revolution, in Paris.

His understanding of mechanical principles and their best proportions combined with the uninhibited and elegant style of his work place him in a category apart from all other horologists. His contributions to the development of the modern watch include the club tooth lever escapement, shock proof suspension for the balance wheel and the terminal curve to the spiral spring.

Buck, D. Azro. Watchmaker of Worcester, Massachusetts. Inventor of the Waterbury revolving movement watch, patented in 1878.

Cole, James Ferguson, 1798-1880. Inventor of many variants of the lever and other escapements. Manufacturer of high grade watches and clocks.

Da Vinci, Leonardo, 1452-1519. Probably best known as the painter of 'The Last Supper' and 'Mona Lisa'. His talent for civil and military engineering, mechanics and biology, coupled with his general philosophy, place him apart as probably the greatest genius of all time. His contribution to horology was the visualization of the pendulum as a controller for clock escapements.

Debaufre, Peter. Patentee with Nicholas Facio in 1704 of a method of piercing precious stones for use as pivot bearings. He also invented an escapement, named after him, employing two 'scape wheels mounted on a common arbor and impulsing directly to the balance staff.

Dennison, Aaron, 1812-1895. Founder with E. Howard and D. P. Davis of watch factory at Roxbury, U.S.A. After the factory failed it was reorganized and became Tracey, Baker & Co. Dennison remained with the company and for a while lived in Zurich, Switzerland, and supervized the manufacture of parts for the company's watches. In 1870 he moved to Birmingham, England, and four years later established a watch case factory there. He continued with this successful venture until his death.

Dent, Edward John, 1790-1853. Celebrated maker of watches, clocks, and chronometers. In 1830 he started partnership with J. R. Arnold. In 1840 started on his own account at 33 Cockspur Street, London. Builder of the famous Westminster clock.

Earnshaw, Thomas, 1749-1829. Inventor of a spring detent chronometer escapement which, being more successful than Arnold's is still used in the manufacture of chronometers. He could not afford the cost of the patent and the money was put up by Thomas Write, a watchmaker, whose name appears on the patent. Earnshaw also devised the superior method of melting the brass onto the steel rim of his compensation balances instead of the then common method of soldering the two metals together. He achieved considerable success as a manufacturer of marine and pocket chronometers and lever watches. Earnshaw repeatedly expressed the greatest contempt for Arnold, whom he publicly accused of pirating his idea of the spring detent. The evidence is that Earnshaw showed his invention to Brockbanks, an associate of Arnold, before it was patented. Arnold then took out a patent for his own spring detent, the idea for which, being different in principle to Earnshaw's, may have occurred to Arnold quite separately but simultaneously.

Emery, Josiah. Born in Switzerland. Settled in England and started in business in London. The first maker of lever watches after Thomas Mudge. He made, between 1782-1795, about thirty watches with this escapement and Arnold's compensation balance and helical spring. Mudge's son declares that the watches were made by Pendleton but they were undoubtedly conceived and financed by Emery and to him must go the considerable credit for their manufacture. Emery also made many fine watches employing the cylinder escapement with ruby cylinder. He died in 1797.

Facio de Duillier, Nicholas, 1664-1753. Born in Switzerland and settled in England in 1687. He patented, with Debaufre, a method of piercing precious stones for use as pivot bearings. The art of jewelling was a closely guarded secret in England until almost the end of the eighteenth century and considerable advantage was gained by this secrecy.

Fasoldt, Charles, 1818-1898. Born in Germany and settled in America in 1849. Celebrated maker of fine clocks and watches employing his patent lever escapement.

Frodsham, Charles, 1810-1871. The most celebrated of the numerous Frodsham family of watchmakers. A prominent maker of very high grade chronometers and watches. He devised, in 1868, a form of electrical contact for taking signals from chronometers.

Galilei, Galileo, 1564-1642. Astronomer, mathematician and optical mechanic. One of the first to construct a telescope for studying the heavens. He established the laws of falling bodies from which he deduced the pendulum to be isochronous.

It was later shown by Huygens to be nearly so only for very small arcs. His drawings for a pendulum clock still exist in the Medicean Archives in Florence. Construction of the clock was started by his son Vincenzio who died in 1649 before it was completed.

Goddard, Luther, 1762-1842. Set up business in Shrewsbury, Massachusetts, in 1809. The business flourished by reason of the favourable conditions created by the Jefferson embargo on imports. He manufactured some 500 watches, until in 1817 he retired to preach the Gospel in Worcester, Massachusetts. The watches, using verge escapement with fusee, were signed at various times L. Goddard and Sons, L. and P. Goddard, and D. P. Goddard and Cie.

Graham, George, 1673-1751. Elected Fellow of the Royal Society in 1721. Inventor of the dead beat escapement for clocks, he also introduced mercurial compensation for his pendulums to maintain the length constant with change of temperature. For his watches he used the cylinder escapement which he developed and introduced in 1726.

Grant, John, died 1810. Maker of fine lever escapement watches using helical springs and compensation balances. His variants of the escapement include the 'chaff cutter' in which the pallets work at right angles to the plane of the 'scape wheel, and a watch with two balances geared together. This latter was an attempt to eliminate variation in the balance amplitude caused by careless handling of the watch in wear.

Harrison, John, 1693-1776. Inventor of the first successful marine timekeeper and winner of the £20,000 reward offered in 1714 by the Act of Queen Anne for a successful method of determining longitude. His first clocks were made almost entirely of wood, which material came readily to hand since his father was a carpenter. His first marine clock also employed wood for the wheels. About 1725 he invented a compensated pendulum composed of alternate rods of steel and brass. This system, called a 'grid iron' relied upon the upward expansion of the brass counteracting the downward expansion of the steel. For his clocks and early marine timekeepers he used the frictionless 'grasshopper' escapement invented by his brother James.

Harwood, John, 1893-1964. English watchmaker and jeweller. Inventor of a mechanism for automatically winding a watch by the motion of the wrist. For his work in the development of the system Harwood was awarded the Gold Medal of the British Horological Institute in 1957.

Hautefeuille, Jean, 1647-1724. Inventor of the rack lever escapement and a form of balance spring in the shape of a straight wire pinned to the pallet staff.

Henlein, Peter, 1479-1542. Master locksmith of Nuremberg. The earliest recorded maker of watches. It is written of him in 1511 that he made watches of iron which both indicated and struck the hours and which could be carried in the purse. There are records of watches ordered from him between 1521 and 1525. There are no surviving examples of his watches.

Hetzel, Max. Co-developer with William Bennet of the Bulova Accutron electronic watch. Born in Switzerland, Hetzel settled in America in 1956 to complete the development of the Accutron. He is now chief physicist of the company.

Hooke, Robert, 1635-1703. First curator of experiments at the Royal Society. Claimant to the invention of the balance spring and the anchor escapement for pendulum clocks. Hooke worked closely with Tompion who did the practical work of his experiments on the application of the balance spring.

Howard, Edward, 1813-1904. In 1842, at the age of 29, he started clock manufacturing at Roxbury, and in 1850 he met A. L. Dennison, and together they started a watch factory which failed in 1857. Howard began again on his own and was successful enough to establish a reputation second to none. His work set the standard for high grade watchmaking in America.

Huygens, Christian, 1629-1695. Eminent physicist, mathematician and astronomer, and the first successfully to apply the pendulum to clocks in 1657 and the balance spring to watches in 1675. His practical contributions to the improvement of timekeeping in clocks and watches are among the most important developments in horology.

Leroux, John, 1744-1808. London maker of fine watches. A watch with lever escapement by him in the Guildhall Museum, London, is hallmarked 1785. The impulse, or lift, is delivered entirely by the 'scape wheel teeth acting on pointed pallets. It is the earliest known watch in which the lever is drawn to the bankings by recoiling the train. This action ensures that the safety dart is quite free of the balance roller during the supplementary arc.

Le Roy, Pierre, 1717-1785. Son of the famous Julien Le Roy [1686-1759] of Paris whose fine work and many improvements started the rise and refinement of French watches. Pierre devoted almost the whole of his time to experiments in the improvement of marine timekeepers. He invented a detached escapement in 1748 and produced a rather unconvincing example of this in 1766. It is still to be seen in the Musée des Arts et Métiers in Paris. In this machine the impulse is delivered at the edge of the balance which is compensated for temperature changes by glass tubes containing mercury.

Litherland, Peter, 1756-1804. Patentee in 1791 of a rack lever escapement. Watches with this escapement, originally the invention of the Abbé de Hautefeuille in 1722, were made in great quantities in Lancashire. Hautefeuille's escapement caused a recoil of the 'scape wheel but Litherland's was dead beat and superior in this respect. Some of his watches employ a train of three wheels only, with 30 teeth to the 'scape wheel. When used with these watches the seconds dial was marked with 15 divisions and the hand rotates four times per minute.

Massey, Edward, 1772-1852. Inventor of a form of detached lever escapement and a means of winding a watch by depressing the pendant. His detached lever, patented in 1815, was the first, simple detached lever since the more complicated work of Mudge, Emery, Leroux and Grant. Its simplicity and good rate encouraged other English makers to take an interest in the lever escapement.

Mudge, Thomas, 1715-1794. Inventor of the detached lever escapement from which all modern lever watches are developed. There is some evidence that he invented the escapement in 1754 but this is not very certain. He had experimented with lever escapements in the 1760s but it was not until 1770 that he succeeded in making a watch with this escapement. The watch was sold to George III and remains in the possession of the Royal Family. In 1771 he moved to Plymouth leaving his business in the hands of William Dutton with whom he had entered into partnership in 1775. He devoted the remainder of his life to the development of his constant force marine timekeeper which was eventually to prove a costly failure.

Pendleton, Richard. Said to have made Emery's lever watches. He made only two known watches bearing his name. He was, with others, later employed by Thomas Mudge junior to make marine timekeepers to his father's plan. Judged by the standard of Emery's watches and the replica Mudge marines he was an exceptionally fine workman. Pendleton died in 1808.

Perigal, Francis. Maker of fine and interesting watches. A lever watch by him is signed 'Watchmaker to the King'. This appointment was conferred upon him in 1784. Little is known of Perigal but the few surviving watches bearing his name are of considerable mechanical interest and are all of fine quality.

Pitkin, Henry, 1811-1846, **James Flagg** 1812-1870 [brothers]. Established in 1838 at Hartford, Connecticut, a watch factory and endeavoured, unsuccessfully, to manufacture the first entirely American watch.

Potter, Albert, 1836-1908. Maker of very high grade watches employing lever and chronometer escapements. In 1876 he moved from America, where he was born, to Switzerland. His work, both in execution and design, is regarded as quite exceptional. In later years, seeing no future in the watch industry for his talents, he turned to other interests. In particular he endeavoured to perfect a steam car, but after several years' work was obliged to admit defeat. Even today in Switzerland his name evokes respect among the few remaining hand-craftsmen.

Prest, Thomas, died 1855. Factory foreman to J. R. Arnold [son of John Arnold] and inventor of a form of keyless wind which was sometimes used by J. R. Arnold in his going barrel lever and cylinder watches.

Quare, Daniel, 1648-1724. Celebrated clock and watchmaker. Inventor of a quarter repeating mechanism for watches, for which in 1686, James II granted a patent, considering it in his opinion to be superior to Barlow's mechanism invented in the same year.

Savage, George, 1767-1845. Inventor in 1815 of a lever escapement known as the Savage two pin. The unlocking is performed by two pins on the balance roller and the impulse is delivered by what would normally be the guard, or safety pin of the conventional English lever. Difficult to make it was not much used but, when properly constructed, had the merit of considerably reducing the unlocking friction and of impulsing the balance at a more advantageous radius. The escapement played an important part in establishing the superiority of the lever for pocket use. In 1818 Savage emigrated to Canada where he established a successful retail business.

Tompion, Thomas, 1639-1713. Celebrated maker of fine clocks and watches.

Although he worked closely with Robert Hooke in his experiments with the balance spring, the two men seem to have had little sympathy for each other. Tompion, whose watches were selling quite well without balance springs and, no doubt, considered they would sell better with Huygens' spiral spring, regarded Hooke as something of a nuisance. Hooke, finding Tompion tardy and unco-operative in making the springs to his pattern refers to him in his diary as a 'slug' and later a 'clownish, churlish Dog'. Tompion devised the form of regulator in which a pinion, when turned by the winding key, moves a curved rack carrying the curb pins.

Plates

1a, b. *Jacques Bulke*, *c.* 1600. English. Verge escapement. Pre-balance spring. Pinned on balance cock. Fusee and gut. 'Bow and arrow' ratchet set up. Decorated border to back plate. Clock watch with hour aperture in count wheel cover. Gilt engraved dial. Single blue steel hand. Note German influence in pierced cover and barrel edged case. Edge of case pierced to make bell more audible.

1a

1b

2a, b. *Hugh Roberts*, *c.* 1620. English. Verge escapement. Pre-balance spring. Pinned on balance cock. Fusee and gut. Decorated border to movement. 24-hour chapter ring on gilt background engraved to show date, month, sign and degree of zodiac, age phase and aspect of the moon. Single blue steel hand and brass pointers for subsidiary information. Gilt metal case with engraved silver band.

3a, b. *Henry Terold*, *c.* 1640. English. Verge escapement. Pre-balance spring. Ratchet set up. Pinned on balance cock. Fusee and gut. Engraved gilt dial with silver chapter ring surrounding the Adoration of the Magi. Single blue steel hand. Gilt edges to the silver case which is engraved with scenes depicting on the front the Annunciation, on the back the Nativity. Inside front the Baptism. Inside back the Crucifixion.

PLATE II

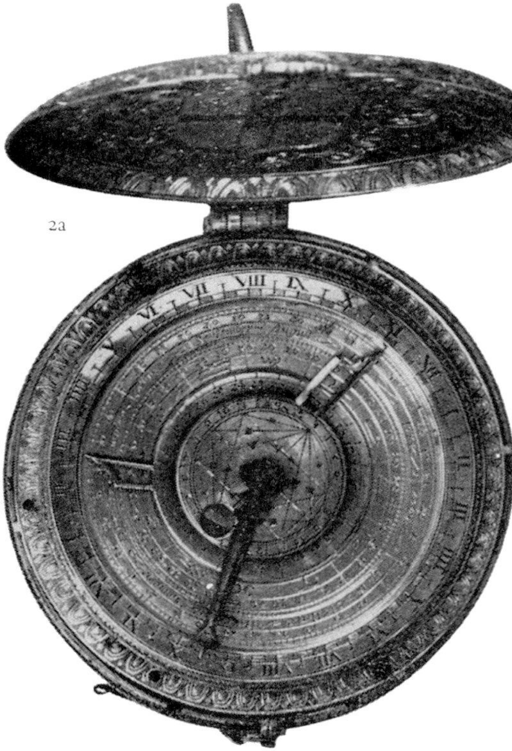

2a

3b

3a

2b

4a, b, c. *Robert Grinkin, c.* 1640. English. Verge. Pre-balance spring. Pinned on balance cock. Fusee and gut. Early worm and wheel set up without index dial. Early square pillars engraved but unpierced. Gilt dial with silver calendar ring and single blue steel hand. Note disappearance of decorated border to movement and pendant drilled left to right instead of earlier fashion of front to back. Silver gilt case with outer, both plain. Watches of this type are called 'puritan'.

5a, b, c. *Edward Clough, c.* 1640. English. Verge. Pre-balance spring. Screwed on balance cock. Early four wheel train going for 30 hours. The extra power is supplied by an additional barrel. Note early long link chain and two worm and wheel set up regulators. In winding the watch the chain from the first barrel is wound onto the fusee. The second barrel is wound at the same time by its chain being drawn onto the first barrel. Silver dial with gilt centre. Two blue steel hands. Silver pair-case, the outer covered in leather and decorated with pin work.

4a

4b

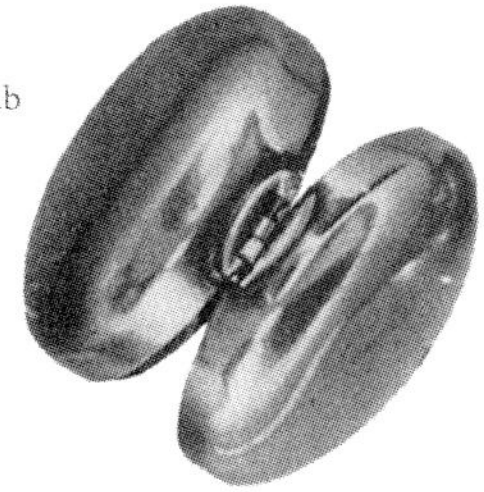

4c

5b

5a

5c

6a, b. *Thomas Tompion*, *c.* 1680. English. Verge with balance spring. Fusee and chain. Tompion regulator with index dial, later universally adopted by English makers. Gilt dial with apertures at top for day of the week, bottom right for month and zodiac sign, bottom left for quarter of day. Silver chapter rings for phase and age of moon, day of the month and hour of the day. The matt effect of the background is achieved by hammering with a rough, sharp edged punch.

7a, b. *Ahasuerus Fromanteel*, *c.* 1695. English. Verge with spring and regulator. Fusee and chain. Silver sun and moon dial (see text, p. 24). Subsidiary dial for seconds quite common to this type of dial. Single blue steel hand for minute indication. Plain silver pair-case.

6a

6b

7a

7b

8. *John Bushman, c.* 1695. English. Verge with spring and regulator. Fusee and chain. Silver wandering hour dial (see text, p. 24) with relief of William III and military appurtenances in gilt. Watches by Bushman are usually decorated in this manner and have, in addition, the Royal Coat of Arms engraved on the balance cock. Silver pair case with *repoussé* decoration to the outer.

9a, b. *Peter Fardoil, c.* 1700. English. Verge with balance spring. Regulator on back plate with arbor passing through to balance under dial and gearing just visible. Fusee and chain. Note increasing decoration of pillars. Gold dial *champlevé* and engraved with aperture to show mock pendulum, in this case carried to extreme absurdity by being inverted. Hours indicated by *fleur de lys* in lower aperture concentric with minute ring and hand. Plain gold pair-case.

10a, b. *Joseph Banks, c.* 1700. English. Verge with spring and regulator. Fusee and chain. Note unbroken edge to cock foot, almost universal by this date. Silver differential dial (see text, p. 24), with revolving chapter ring surrounded by military motif in relief. Single blue steel hand. Plain silver pair-case.

PLATE V

9a

9b

8

10a

10b

11a, b. *William Peacock*, *c.* 1700. English. Verge escapement with balance spring and regulator. Silver six hour dial (see text, p. 24), with gilt surround. Single blue steel hand. Silver pair-case, the outer covered with tortoiseshell and inlaid with silver.

12a, b. *George Graham*, *c.* 1735. English. Cylinder escapement. Gold *champlevé* dial with blue steel hands. Gold pair-case, the outer chiselled and engraved. Note the curved ends to the case hinge succeeding the earlier plain square ends and diamond and stone to balance cock.

11a

11b

12a

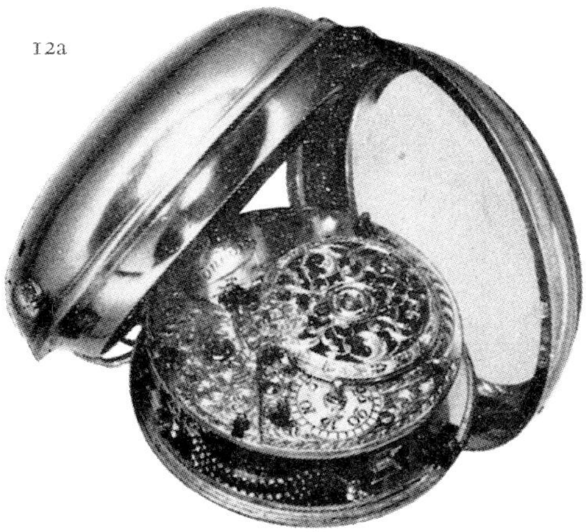

12b

13a, b. *Larcum Kendal*, h.m. 1777. English. Pivoted detent escapement. Regulation by altering the weight of the balance. Temperature compensation by spiral bimetallic strip and pivoting curb pins. Fusee and chain with Harrison's maintaining power to prevent reversal of train during winding. Enamel dial of regulator pattern used later by Emery and Grant for lever watches. Blue steel hands. Plain silver case. Wound through shuttered aperture at back.

14a, b. *Thomas Martin*, *c.* 1778. English. Ruby cylinder escapement. Fusee and chain. Enamel dial with outer seconds divisions. Subsidiary dials for mean time at top, minutes at left, regulation at right and $\frac{1}{4}$ seconds at bottom. The revolutions of the centre seconds hand are recorded by the minute dial. Gold hands. Gold case with translucent enamel over engine turning and chain of gold circlets around bezel and edge of back. Oval enamel plaque to centre of back surrounded by white line edged with gold.

13a

13b

14a

14b

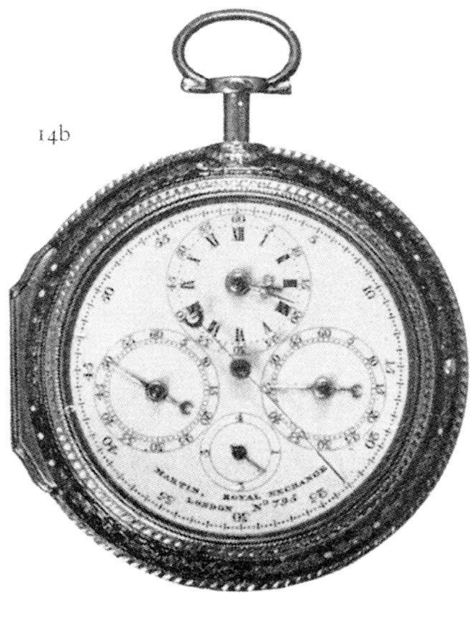

15a, b. *John Arnold*, No. 37, h.m. 1779. English. Pivoted detent escapement. Fusee and chain with maintaining power always used with precision watches. Helical balance spring with terminal curves. Compensation balance of Arnold's first type for watches with weights moved radially by bimetallic strips. White enamel dial, blue steel hands. Silver, engraved consular case.

16a, b. *John Arnold*, No. 9/61, h.m. 1789. English. Pivoted detent escapement. Helical gold spring with terminal curves. Compensation balance of Arnold's second type with the longer bimetallic strips folded into two loops each for compactness. Fusee and chain. $\frac{1}{4}$ repeat, the hammers hitting blocks in the case. This method is called in England dumb repeating. In this example a button in the bezel, if pressed, prevents the hammers touching the case and allows the blows to be felt and not heard. Enamel dial, gold hands. Gold consular type case with press in pendant for repeat.

17a, b. *John Arnold*, No. 109/410, h.m. 1784. English. Spring detent escapement. Gold helical balance spring. Compensation by Arnold's third type of balance with bimetallic strips and adjustable weights screwed to a plain balance wheel. Fusee and chain. Enamel dial. Blue steel hands. Plain gold consular case.

18. *John Arnold*, No. 145. 1800. English. Spring detent escapement. Gold helical balance spring. Compensation balance of Earnshaw's type but retaining the screwed adjusting weights of Arnold. This (movement only illustrated) represents the final form of Arnold's pocket chronometer made in his lifetime.

PLATE VIII

15a

15b

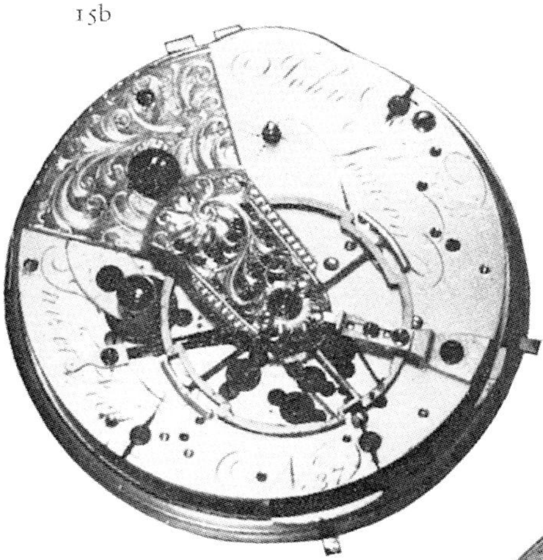

16a

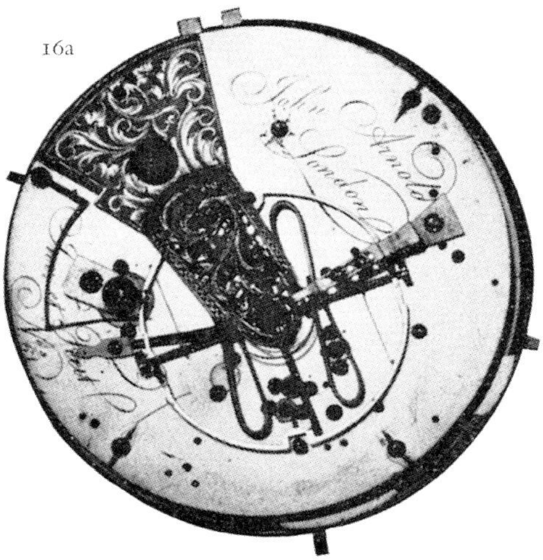

16b

17b

18

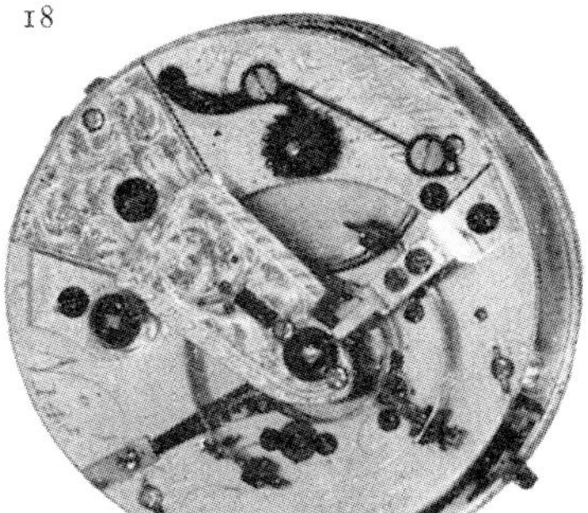

19a, b. *Peter Litherland*, h.m. 1794. English. Rack lever escapement. Regulation by index and scale on balance cock. This method of effectively varying the length of the balance spring was commonly used on non-precision watches. Fusee and chain. Enamel dial. Gold hands. Plain silver pair-case.

20a, b. *Mudge & Dutton*, h.m. 1800. English. Lever escapement. Helical balance spring. Compensation balance. Fusee and chain. In all outward appearances this movement is similar to Emery's levers and, like them, was probably made by Pendelton. Enamel dial and gold hands typical of Mudge watches of the period. Plain gold case. $\frac{1}{4}$ repeat by pressing in pendant.

19a

19b

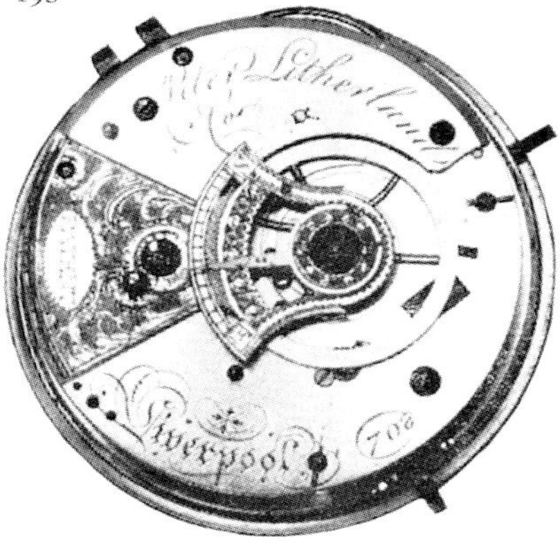

20a

20b

21a, b. *Thomas Earnshaw*, *c.* 1798. English. Spring detent escapement. Spiral balance spring. Compensation by bimetallic curbs arranged to reduce the freedom of the spring between their tips for increase of temperature and vice versa for a reduction in temperature. Fusee and chain. Enamel dial. Blue steel hands. This form of compensation was used by Earnshaw to reduce the weight of the balance and thus increase its resistance to damage by rough usage. Mean time regulation is effected by sliding the carrier for the curbs (in this form called sugar tongs) along the scale engraved at the edge of the plate. White enamel dial. Gold hands. Plain silver pair-case.

22a, b. *Thomas Earnshaw*, *c.* 1800. English. Spring detent escapement. Spiral balance spring. Compensation balance with Earnshaw's weights. Small pins in side of weights locate in holes in spokes radiating from centre of balance to prevent distortion of balance rims. Fusee and chain. Enamel dial. Gold hands. Plain gold pair-case.

21a

21b

22a

22b

23a, b. *Edward Massey*, h.m. 1819. English. Massey lever escapement. Spiral balance spring with regulator. Going barrel. Mainspring wound by pumping the pendant, also a Massey patent. Enamel dial with subsidiary seconds signed 'Edwd. Massey Improved Escapement'. Blue steel hands. Plain silver case.

24a, b. *William Ilbery*, *c.* 1820. English. Duplex escapement. Going barrel. Enamel dial. Gold case with enamel domestic scene on back. Bezels and pendant set with half pearls. Edge of case enamelled blue, red and white, and set with half pearls. Ilbery was famous for his watches especially developed for the Chinese market. They were freely copied by other makers for export to China, but movements signed by Ilbery are not common.

PLATE XI

23a

23b

24a

24b

25a, b. *J. R. Arnold*, h.m. 1820. English. Ruby cylinder escapement. Spiral balance spring. Compensation curb on regulator. Going barrel. Prest keyless wind. Enamel dial. Gold hands set by pushing with the finger. Plain gold case.

26a, b, c. *J. F. Cole*, 1830. English. Lever escapement. Spiral spring with terminal curve and regulator. Compensation balance with long screw heads typical of Cole's balances. Gold engine turned dial with subsidiary dials for hours, minutes and seconds. Gold engine turned case with à tact hand and touchpieces at band of case. With this device, invented by Breguet, a close approximation of the time can be had by turning the hand until stopped by the internal mechanism and then feeling its position relative to the touch pieces placed at the twelve hour positions on the case band.

PLATE XII

25a

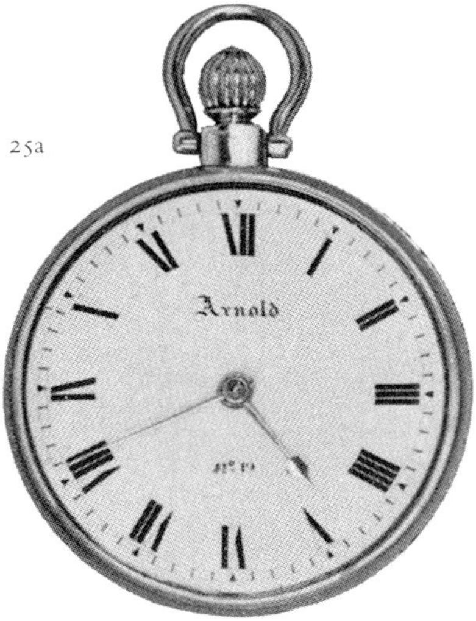

25b

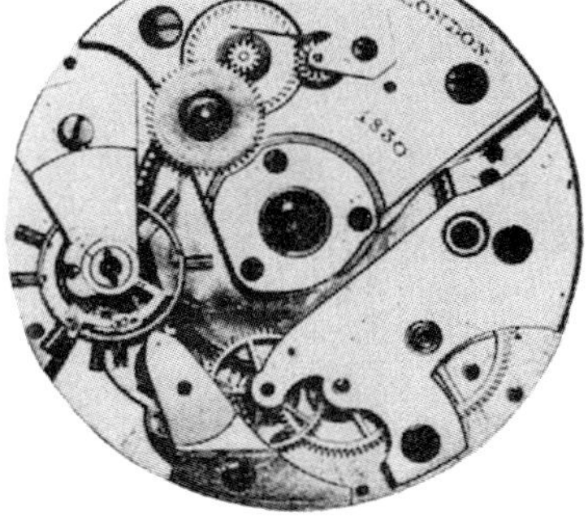

26b

26a

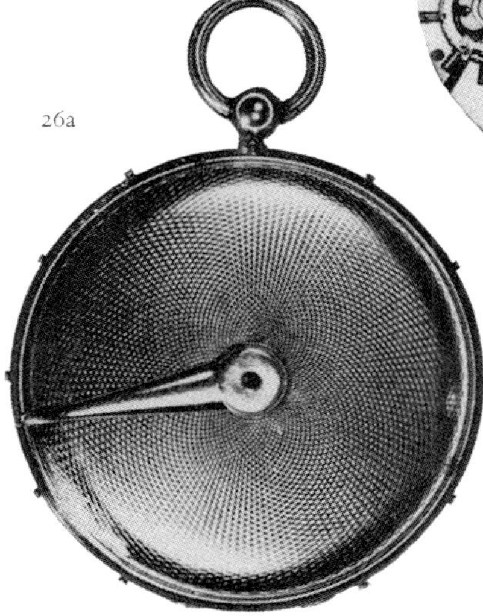

26c

27a, b. *Thos. Penkett.* h.m. 1832. English. Rack lever escapement. Spiral spring with regulator. Going barrel. Mainspring wound by pedometer weight. This system of winding originated with A. L. Perrelet (Swiss) and was brought to perfection by A. L. Breguet. Enamel dial. Gold hands. Plain silver case.

28a, b. *Parkinson & Frodsham, c.* 1860. English. Chronometer escapement. Compensation balance with helical steel spring and end curves. Fusee and chain. White enamel dial with blue steel hands. Subsidiary dial for up and down indicator. Plain silver hunting case.

PLATE XIII

27a

27b

28a

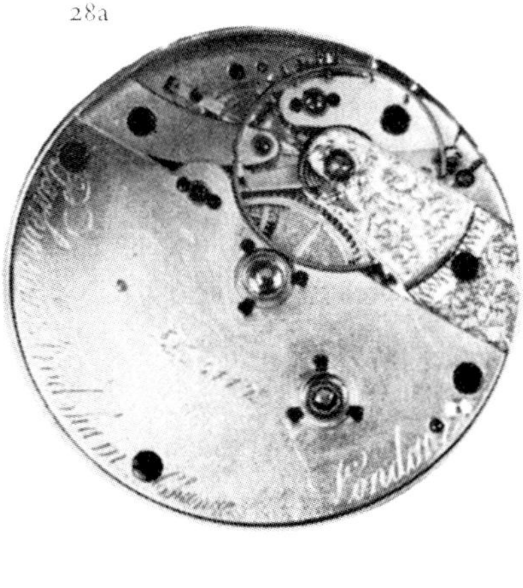

28a

29a, b. *W. M. Godfrey*, h.m. 1870. English. Lever escapement. Spiral balance spring with terminal curve and regulator. Compensation balance. Fusee and chain. Enamel dial with subsidiary dials for seconds and up and down to indicate hours run. Blue steel hands. Plain gold case, the front released by pressing the button in the pendant. This represents the final development of the English key wound lever watch.

30a, b. *John Bull* (Lancashire Watch Co.), *c.* 1905. English. Pin lever escapement. Spiral balance spring with regulator. Going barrel. Keyless wind. Enamel dial with subsidiary seconds. Blue steel hands. Gilt brass case. A successful attempt to make a mass produced watch in England which failed through the financial instability of the company.

29a

29b

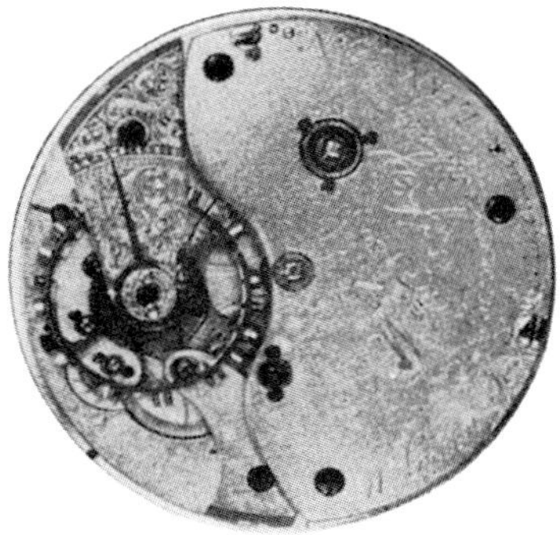

30a

30b

31a, b. *Chas. Frodsham, c.* 1910. English. Lever escapement. Spiral balance spring with terminal curve, free sprung. One minute tourbillon. Compensation balance. Fusee and chain. Keyless winding. Hand set by pushing button at one o'clock and turning crown. Metal dial with silver finish. Subsidiary dials for seconds and up-down. These escapements by Nicole Nielson were fitted to the watches of Frodsham and Dent in addition to those bearing his own name. Plain gold case.

32a, b. *Nicole Nielson, c.* 1913. English. Lever escapement. Spiral spring with terminal curve and regulator. Going barrel. Keyless wind. Enamel dial. Blue steel hands. The centre seconds hand can, by means of the buttons at eleven and one o'clock, be started, stopped and returned to zero. The subsidiary dial at two o'clock recor ds the revolutions of the seconds hand.

PLATE XV

31a

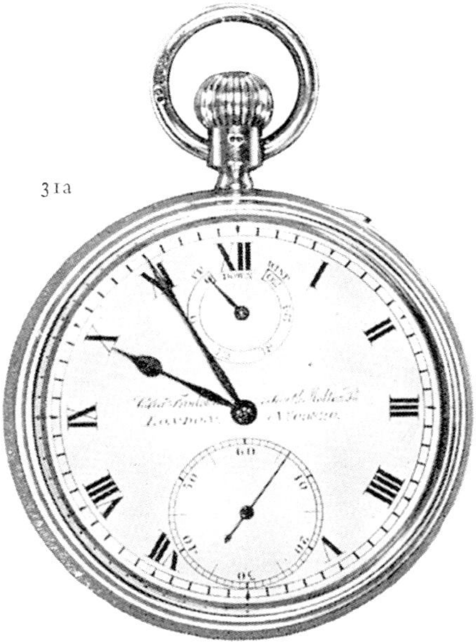

31b

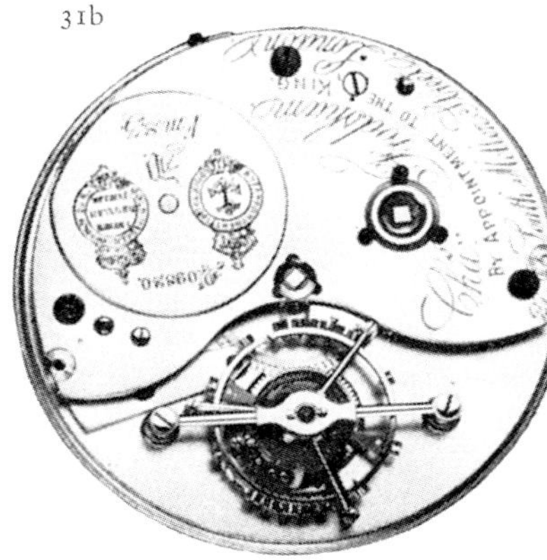

32a

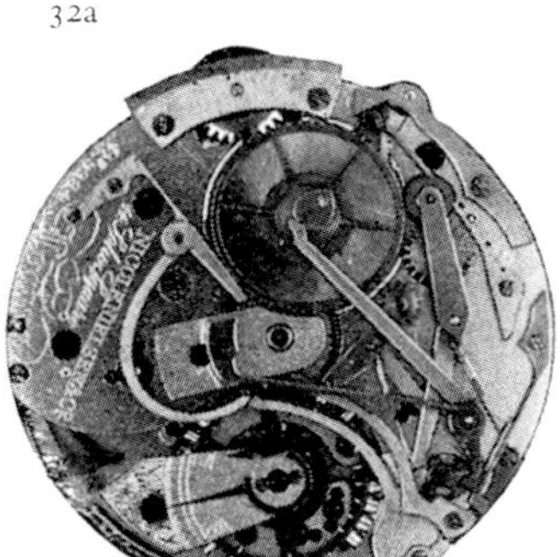

32b

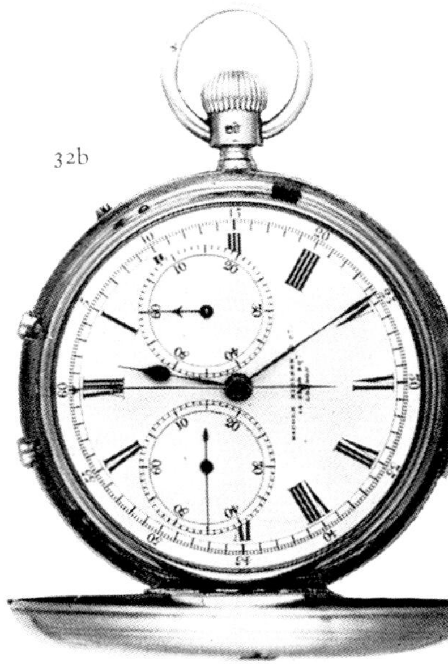

33a, b. *Rotherham*, 1923. English. Lever escapement. Karrusel with 52.5 minute rotation (Bonniksen's patent 1894). Spiral spring with end curve and regulator. Going barrel. Enamel dial. Blue steel hands. Plain gold case.

34a, b. *John Harwood*, *c.* 1930. English. Lever escapement. Compensation balance. Spiral spring with regulator. Automatic winding by oscillating weight seen pivoted at centre of movement. The hands are set to time by rotating the milled bezel. Gold case. Silvered metal dial with luminous hands. (*Illustrated twice original size and without wrist strap.*)

33a

33b

34a

34b

35a, b. *Luther Goddard, c.* 1816. American. Verge escapement. Plain steel balance. Spiral steel spring with regulator. Fusee and chain. White enamel dial with gold hands. Plain silver pair-case.

36a, b. *Dennison, Howard and Davis*, 1854. American. Lever escapement, uncompensated balance. Going barrel. White enamel dial with blue steel hands. Hunting case, the front opened by pressing the knurled button in the pendant.

35a

35b

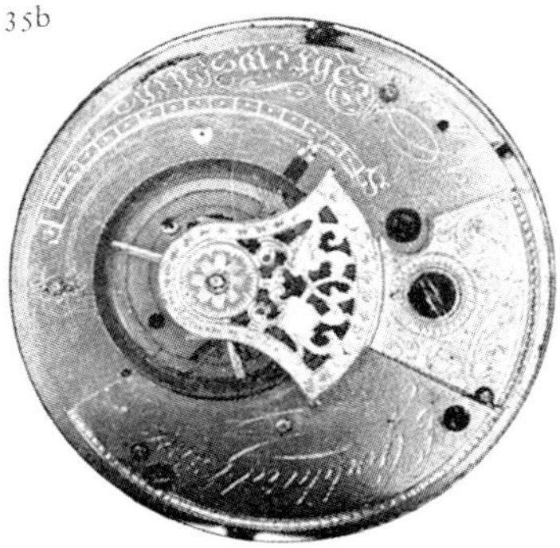

36a

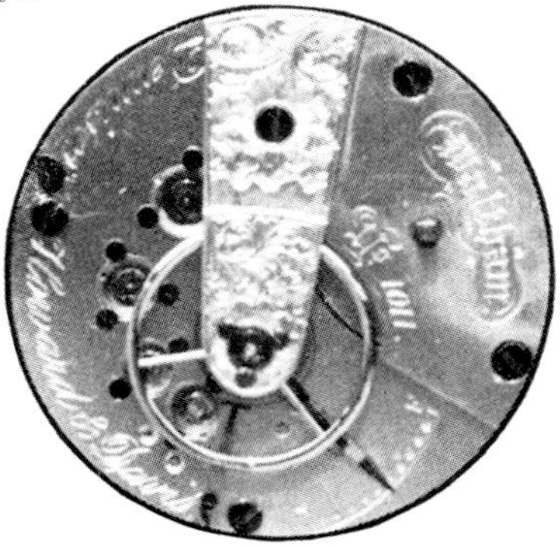

36b

37a, b. *E. Howard*, *c.* 1858. American. 18 size. Lever escapement. Spiral steel spring. Compensation balance. Going barrel with stop work. White enamel dial and blue steel hands. Hunting case, the front opened by pressing the knurled button in the pendant.

38a, b. *E. Howard*, *c.* 1865. American. Size N. Lever escapement. Spiral steel spring. Compensation balance. G. P. Reed's stationary barrel employing maintaining power and stop work. With this arrangement damage to the great wheel teeth in the event of mainspring failure is eliminated. Silver case.

37a

37b

38a

38b

39a, b. *Charles Fasoldt*, *c.* 1865. American. Fasoldt double wheel lever escapement. Spiral steel spring with terminal curve and regulator. Compensation balance. Going barrel. Keyless wind. Hand set by pull up crown. Enamel dial with subsidiary dial for seconds. Blue steel hands. Gold case.

40a, b. *Auburndale Rotary*, *c.* 1876. American. Lever escapement with pointed pallets. Spiral steel spring with regulator. Going barrel carrying train and escapement rotating once in 2.5 hours with internally toothed ring to drive third pinion. Push in crown and turn to set hands. Enamel dial. Blue steel hands. Nickel case.

39a

39b

40a

40b

41a, b. *Albert Potter*, made in Switzerland *c.* 1877. American. Pivoted detent chronometer. Compensation balance. Helical steel balance spring with terminal curves. Patent safety barrel, keyless wind. White enamel dial with blue steel hands. Plain gold case. The escapement used in this watch was invented by Potter in 1875 and had the merit of reducing the unlocking friction by increasing the active radius of the passing spring.

42. *Waterbury Type* A, *c.* 1878. American. Duplex escapement. Spiral spring and regulator. Mainspring in case with rotating arbor carrying movement which turns once per hour and carries minute hand. Glazed paper dial. Blue steel hands. Nickel case. Revolving movements were abandoned with Series E, *c.* 1882.

43. *Waterbury Type* F, *c.* 1888. American. Duplex escapement. Spiral steel spring with regulator. Uncut compensation balance. Paper dial. Blue steel hands. Nickel case. Movement only illustrated.

PLATE XX

41a

41b

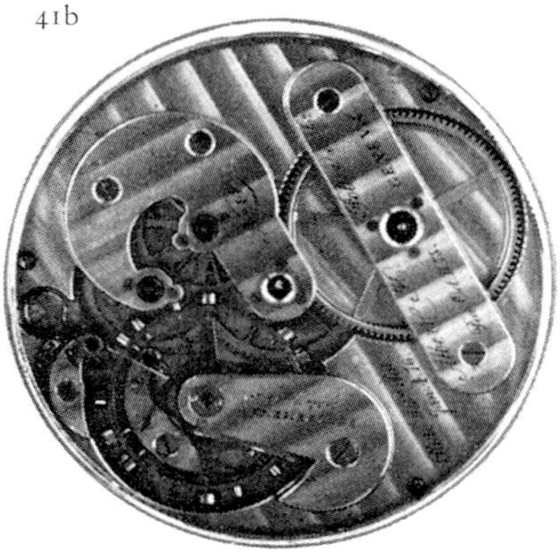

42

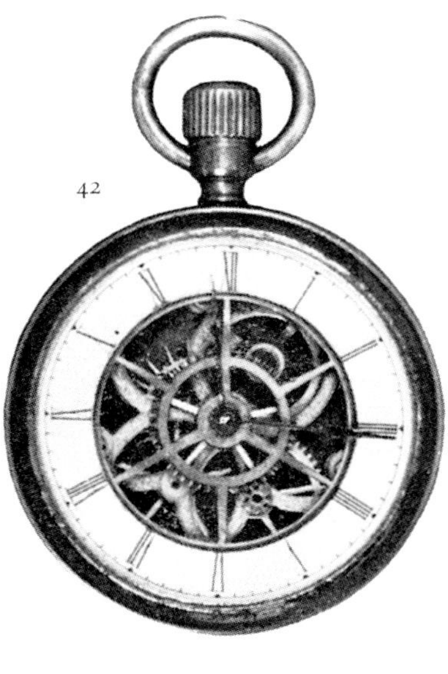

43

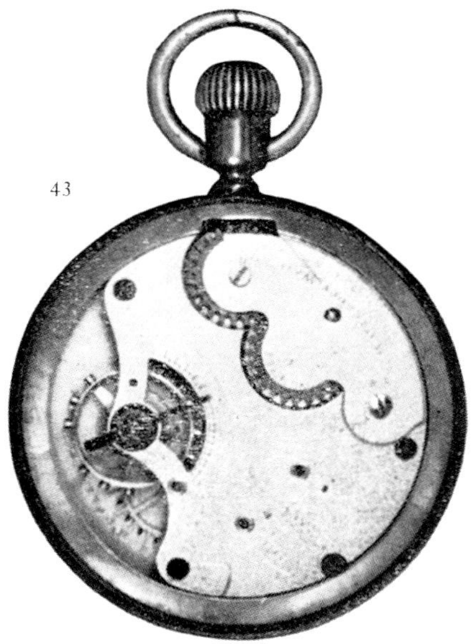

44a, b. *Waterbury Type* I, *The Trump*, *c.* 1889. American. Duplex escapement. Spiral spring with regulator. Plain balance. Going barrel. Enamel dial with subsidiary seconds. Blue steel hands of superior finish. Nickel case. This watch, together with models A and F are representative of the complete range of the Waterbury watches.

45a, b. *Waltham*, *c.* 1890. American. Lever escapement. Spiral steel balance spring without regulator. Going barrel. The back plate and balance cock are made from crystal allowing the whole of the mechanism to be seen. Enamel dial. Blue steel hands. Gold filled case.

44a

44b

45a

45b

46a, b. *Ingersoll* (manufactured by New Haven Clock Co.), *c.* 1892. American. Pin lever escapement. Spiral spring with regulator. Going barrel wound by folding key. Hand set by knurled button at centre. Paper dial with subsidiary dial for seconds. Black painted hands. Gilt metal case with false winding crown.

47a, b. *Waltham Premier Maximus*, *c.* 1909. American. Lever escapement. Spiral spring with terminal curve and regulator with micrometer adjustment. Compensation balance. 23 jewel movement. White enamel dial and blue steel hands. Subsidiary sector to indicate state of wind of mainspring. Gold case.

46a

46b

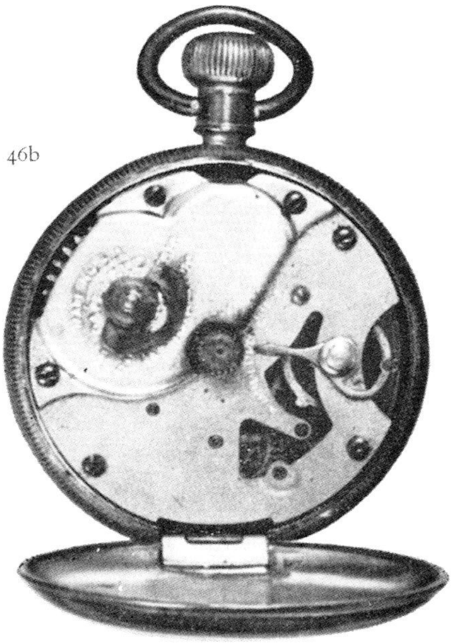

47a

47b

48a, b. *E. Howard, c.* 1909. American. Lever escapement. Compensation balance. Spiral steel balance spring with terminal curve and regulator. Patent safety barrel with keyless wind. 23 jewel movement. White enamel dial with blue steel hands. Plain gold case.

49a, b. *Hamilton*, 1966. American. The world's first electric watch, introduced in 1957. Monometallic balance wheel with special alloy balance spring designed to compensate for temperature changes and resist residual effects of magnetism. Impulse by magnetic field generated in coil of wire attached to balance. Battery power source can be seen to left to balance. Cases and dials can be obtained in a variety of modern styles and materials. This example with gold case and hands. Silver dial with gold numerals and baton quarters.

50a, b. *Bulova, Accutron*, 1966. American. The world's first electronic watch. Introduced in 1960. Battery power supply is seen at the right of the movement. The electronic circuit is contained in the dark irregular shaped moulding seen at the left. The tuning fork, secured to the edge of the movement behind the back-plate, terminates in the two cups containing permanent magnets and enclosing the energizing coils. Cases and dials available in a wide range of modern styles. This example with stainless steel case and black dial with stainless numerals and quarter batons. Arrow-shaped hand to indicate time on 24-hour outer dial. (*Illustrated* 1½ *times original size and without wrist strap.*)

48a

48b

49a

49b

50a

50b

INDEX

Accutron electronic watch, Bulova, 56, 69, 73
American Horologe Company, 48
Arnold, J., 27, 28, 33, 34, 35, 36, 37, 43, 65, 66, 69, 75
Arnold, J. R., 35, 43, 69, 70, 71, 75, 100
Auburndale Rotary watch, 50
Auburndale Watch Company, 50

Balance, 30, 34, 56, 59,
  arbor, 30, 31, 39
  cock, 18, 19,
  compensation, 39, 49
  spring, 23, 24, 29, 60, 61, 67, 73
    helical, 33, 34, 43, 69, 71, 72
    spiral, 22, 49, 70
    terminal curve, 35, 69, 70
  weights, 34
  wheel, 60
Banking, 31, 38
Barlow, E., 27, 69, 75
Barrel, 43
  safety, 49
  stationary, 49
Battery, 56
Benedict, C., 51
Benedict & Burnham Manufacturing Company, 50
Bennet, W., 57, 69, 73
Bimetallic curb, 33
Bimetallic strips, 34, 35
Boston Watch Company, 48
Brass, 16, 18, 28, 72
Breguet, A. L., 39, 42, 43, 45, 55, 67, 68, 69
British Horological Institute, 73
Bruhl, Count von, 37
Buck, D. A., 50, 70
Bull, J., 104
Bulova, A., 56
Bulova, J., 56
Bulova Watch Company, 56, 57, 69

Capacitor, 56
Carey, J., 47
Cases:
  Barleycorn engine turning, 43
  crucifix, 19
  crystal, 19
  drum shaped, 15
  enamel, 33
  fish-skin, 20, 23
  gilt metal, 15, 19
  gold, 19, 23, 36, 43, 49, 51
  leather, 20, 23, 28
  octagonal, 19
  outer, 20
  oval, 19
  pair, 28, 33
  'puritan', 20
  répousse, 28
  silver, 19, 23, 36, 43, 51
  spherical, 15
  tortoiseshell, 20
Chain, 23
Charmilles, 53
Clocks, 46
  regulator, 52, 53
  spring-driven, 15, 16, 20
  weight-driven, 15, 21
Cole, J. F., 44, 49, 70, 100
Colt, W. D., 50
Compensation, 31, 34, 35
  'grid iron', 72
  mercurial, 26, 28, 72
Constant force, 53
Contrate wheel, 59
Crown wheel, 59
Curb pins, 29
Curtis, S., 48

Da Vinci, L., 21, 70
Davis, D. P., 48, 70
Debaufre, P., 27, 70, 71
Dennison, A. L., 47, 48, 70, 73
Dent, E. J., 44, 70

Detent, 35, 65
  spring, 35, 71
Dials:
  champlevé, 27
  differential, 24
  enamel, 27, 36, 43
  moon, 18
  paper, 51
  revolving hour, 24
  six hour dial, 24
  sun and moon, 24
  zodiac, 18
Dijon, 17
Dividing engine, 22
Dutton, W., 38, 74

Earnshaw, T., 35, 36, 43, 65, 70, 71
Ebauchés, 47
Elasticity, 29
Electric watch, 55
Electro-magnetic, 56
Electronic, 56
Elgin Watch Company, 51
Emery, J., 37, 38, 39, 71, 74, 75
Escapements:
  chronometer, 35, 36, 37, 38, 41, 43, 53, 56, **65-66, Fig. 4, Fig. 5,** 67, 68, 69, 70, 71, 75
  club tooth lever, 67, 70
  cylinder, 26, 27, 33, 41, 44, **61-62, Fig. 2,** 63, 66, 71, 72, 75
  Debaufre's escapement, 70
  detached, 33
  divided lift, 49
  duplex, 39, 40, 41, 43, 50, **63-64, Fig. 3**
  Fasoldt's escapement, 52, 71
  Graham's escapement. *See* cylinder
  'grasshopper', 28, 72
  horizontal. *See* cylinder
  lever, 30, 31, 32, 37, 40, 42, 43, 49, 53, **67-68, Fig. 6, Fig. 7,** 70, 71, 72, 73, 74 75
  Massey's escapement, 41, 42
  rack lever, 41, 42, 73, 74
  resilient, 49
  right angle lever, 38
  'straight line lever', 38, 39
  'tick-tack', 26
  'two pin' lever, 42, 75
  verge, 21, 26, **59-60, Fig. 1,** 61, 62, 66, 72

Facio de Duillier, N., 27, 70, 71
Fasoldt, C., 52, 53, 71
Foliots, 59
Fork, 30, 42
  vibrating, 56
French watchmaking, 16, 17, 18, 19, 27, 39
Frodsham, C., 44, 71, 102
Fusee, 16, 72

Galilei, G., 21, 71
Galilei, V., 21, 72
George III, 33, 74
German watchmaking, 16, 17, 18, 19
Glasses, 20
Goddard, L., 47, 72
Going barrel, 43
Gould, R. T., 29n., 36n.
Grant, J., 38, 39, 72, 74
Great wheel, 29, 43
Greenwich, 34
Guard pin, 42
Guildhall Museum, 73
Gut, 23
Gyration, radius of, 30

Hamilton Company, 51, 52, 55, 69
Hands:
  concentric, 27
  expanding minute hand 24
  hour hand, 23
  minute hand, 23, 24
  seconds hand, 23, 74
  single, 19, 20, 24
Harrison, James, 28, 72
Harrison, John, 28, 29, 33, 34, 49, 72
Harwood, J., 44, 45, 55, 72, 73
Hautefeuille, J., 30, 41, 73, 74
Henlein, P., 15, 73
Hetzel, M., 57, 69, 73
Hooke, R., 22, 73, 76
Hopkins, J., 50
Howard, E., 47, 48, 49, 70, 73, 110

Howard, E., Watch & Clock Company, 48
Howard Clock & Watch Company, 48, 49
Humidity, 32
Huygens, C., 21, 22, 72, 73

Illinois Watch Company, 51
Impulse, 26, 38, 42, 56, 61, 74, 75
  coil, 56
  cyloidal, 66
  jewel, 65
  pallet, 63
  plane, 67
  radius, 62, 66
  roller, 68
  springs, 29
Inertia, 34
Ingersoll, C. H., 51
Ingersoll, R. H., 51
Ingersoll Watch Company, 50, 51
Iron, 16, 73
Isochronal pin, 32

James II, 69, 75
Jewels, 27, 28, 29, 42, 52, 53, 57, 63, 65, 70, 71
  indexing, 57
  pin, 42
  slotted, 63

Karrusel, 44

Lamination, 29

Le Roy, J., 74
Le Roy, P., 30, 34, 35, 74
Le Roy et Cie, 31
Leroux, J., 38, 39, 42, 73, 74
Lever, 31, 35, 39, 41, 42, 49, 53, 73
  pivoted, 34
Litherland, P., 41, 74
Locke, E., 50
Locking stone, 65, 66
Longitude, Board of, 28, 33, 36
Lyons, 17

Magnet, 56
Mainspring, 16, 23
Maintaining power, 29, 49
Marine chronometer, 17, 28, 30, 31, 34, 52, 69, 72, 74
Marine Chronometer, The, 29n., 36n.
Mass production, 13, 41, 44, 48, 50, 51, 52
Massey, E., 41, 42
Medicean Archives, 21, 72
Mercer, T., 35
Merrit, G., 50
Mudge, T., 28, 29, 30, 31, 32, 33, 37, 38, 71, 74, 75
Mudge, T., Junior, 38, 71, 74
Musée des Arts et Métiers, 74

New England Watch Company, 51
Numerals:
  Arabic, 23
  Roman, 23

Oil, 27, 32, 53, 66

Pallets, 29, 38, 39, 41, 72, 73
  stones, 43, 67
Paris, 17
Pawl, 29
Pendant, 27, 74
Pendleton, R., 38, 71, 75
Pendulum, 21, 59
  mock, 25
Perigal, S., 38, 75
Perrelet, A. L., 55
Pitkin, H., 47, 75
Pitkin, J. F., 47, 75
Plate:
  three-quarter, 39, 49
  full, 48
Potter, A., 53, 75
Potter, W., 53
Potter Bros., 53
Prest, T., 43, 75

Quare, D., 27, 69, 75
Queen Anne, 28, 72

Rack, 30
Ratchet wheel, 29

Rate, 27
Recoil, 26, 38, 59, 61
Regulation, 23
Repeater, 18, 27, 28, 53, 69, 73, 75
Resistor, 56
Roller, 42
Rotating movement, 50, 51
Rouen, 17
Royal Society, The, 22, 72, 73
Ruby cylinder, 28, 33, 71

Safety dart, 38, 39, 67, 73
Savage, G., 42, 75
Schild, A., S. A., 45
Smith's, 45
Split seconds, 50
Stackfreed, 16, 17
Stop watch, 50
Stop work, 43
Supplementary arc, 27, 38, 59, 61, 73
Swiss watchmaking, 44, 45, 49, 51, 55

Temperature errors, 30
Thermometer (Auburndale), 50
Tompion, T., 22, 26, 73, 75, 76, 84
Tourbillon, 44
Tracey & Baker, 48
Tracey Baker & Company, 49, 70
Trains:
  four-wheel, 18, 23
  three-wheel, 18, 74
Transistor, 56
Tyrer, T., 39

Unlocking, 30, 31, 42

Vibration, 30
Viner, 44

Waltham Riverside Maximus watch, 49
Waltham Watch Company, 49, 51
Warren Manufacturing Company, 48
Waterbury Watch Company, 50, 51, 70
Willard, A., 47
Winding, 23
  automatic, 44, 45, 55, 72
  key, 19, 36, 49, 76
  keyless, 43, 49, 75
Worm and wheel, 19, 23
Write, T., 70

BIBLIOGRAPHY

Britten, F. *Old Clocks and Watches and their Makers* 7th edn. Spon, London 1956.

Chamberlain, R. *It's About Time* William Bird Press Inc., New York 1941.
Holland Press Ltd., London 1964.

Chapuis, A. and Jacquet, E. *The History of the Self-Winding Watch, 1770-1931* Batsford, London 1957.

Clutton, C. and Daniels, G. *Watches* Batsford, London 1965.

Gould, R. T. *The Marine Chronometer* Holland Press Ltd., London 1960.

Victoria & Albert Museum *English Watches* H.M.S.O., London 1956.